168 Current Topics in Microbiology and Immunology

WILLIAM S. MASON
CHRISTOPH SEEGER
Fox Chase Cancer Center
7701 Burholme Ave.
Philadelphia, PA 19111
USA

Cover design by courtesy of Dr. Allison Jilbert, Institute of Medical and Veterinary Science, Adelaide, South Australia.

ISBN 3-540-53060-6 Springer-Verlag Berlin Heidelberg New York
ISBN 0-387-53060-6 Springer-Verlag New York Berlin Heidelberg

Library of Congress Catalog Card Number 15-12910
Printed in Germany

Typesetting: Thomson Press (India) Ltd, New Delhi;
Offsetprinting: Saladruck, Berlin; Bookbinding: B. Helm, Berlin.
23/3020-543210—Printed on acid-free paper.

Preface

The discovery of animal models and the possibility of expressing virus in tissue culture cells have catapulted hepadnavirus research from auspicious beginnings more than two decades ago into a dynamic field that has in a short time uncovered the major problems pertinent to the viral replication cycle. However, human hepatitis B virus (HBV) still infects more than 200 million people worldwide and remains one of the most severe human pathogens, despite the availability of efficient vaccines to prevent primary infection. This is a consequence both of the ability of the virus to cause chronic, life-long infections and of the failure of conventional therapies to block virus production and effect a permanent cure of the infected liver. The goal of curing a chronic carrier may, however, be accomplished in the near future as the molecular principles of hepadnavirus replication continue to unfold. It is the purpose of this volume to summarize the results of the latest efforts by investigators in this field and to provide perspectives, emerging from today's knowledge, for the years to come.

The book commences with a chapter by Israel and London describing the consequences of HBV infection on anatomy and function of the liver. One of the major characteristics of hepadnaviruses is their marked hepatoropism. This property is, in fact, believed to be imprinted at least on one level by sequence elements encoding viral promoters and enhancers. Schaller and Fischer discuss how viral gene expression is regulated and how the mechanism of gene expression controls virus production in the infected hepatocyte. The next two chapters deal with the mechanism of viral replication in the liver. Investigations of the mechanism of viral DNA synthesis have long been hampered by the failure to propagate hepadnaviruses in cultured cells. The development of tissue culture systems for the expression of infectious virus has energized a genetic and functional analysis of the hepadnavirus genome. The results from these investigations and current models explaining how hepadnaviruses synthesize DNA from RNA are summarized in the chapter by Seeger et al. Biogenesis of the viral particle is the subject of the next chapter by Ganem, who describes the strategy employed for the synthesis of the viral proteins and discusses the current knowledge of the signals controlling assembly of infectious virus. The following two chapters are devoted to the most severe consequence of hepatitis B virus infection, the development of hepatocellular carcinoma. To date, much of our understanding of the process of hepatocarcinogenesis has been based upon analogies to other systems. Accordingly, Chisari discusses how transgenic mouse models have been used to test some novel ideas about the causes of liver cancer in an infected host. Rogler then describes what is known about the role of hepadnaviruses in hepatocarcinogenesis in their indigenous hosts and discusses current molecular models of liver cancer ontogeny in this setting. An important clue for unraveling the parameters commanding viral gene expression may emerge from the molecular analysis of hepatitis delta virus (HDV), discussed by Taylor. HDV replication, while itself dependent on one or several HBV products, suppresses replication of hepadnaviruses by an as yet unknown process. The closing two chapters of this volume provide a perspective of the long-term goals of hepadnavirus research; the first, by Marion, considers central questions about the development of antiviral therapies for chronic infection, and the second, by Horwich, the possibility of using hepadnaviruses as vectors for liver-directed gene therapy in inherited diseases.

William S. Mason
Christoph Seeger

List of Contents

List of Contributors

(You will find the authors addresses at the beginning of their respective contributions)

Liver Structure, Function, and Anatomy: Effects of Hepatitis B Virus

J. ISRAEL[1] and W. T. LONDON[2]

Hepatitis B virus (HBV) is a common human pathogen that causes extensive morbidity and mortality in human populations throughout the world. Although the virus has been identified in extrahepatic tissues, such as vascular endothelium, bile duct epithelium, bone marrow, and peripheral blood lymphocytes (BLUM et al. 1983), the liver is the major and most important site of infection. It is the disruption of the normal anatomy and functions of the liver that leads to a set of clinically and pathologically defined acute and chronic diseases, including acute and chronic hepatitis and cirrhosis. Chronic infection with HBV is also associated with the development of hepatocellular carcinoma, particularly in the setting of a cirrhotic liver. None of these diseases is specific to HBV infection;

[1] Gastrointestinal Section, Department of Medicine, University of Pennsylvania, Philadelphia, PA 19104, USA
[2] Division of Population Sciences, Fox Chase Cancer Center, 7701 Burholme Avenue, Philadelphia, PA 19111, USA

Current Topics in Microbiology and Immunology, Vol. 168

each can be caused by other infectious and hepatotoxic agents. What is distinctive about HBV is that it can cause all of these diseases as well as no apparent disease. It is this variability in outcome that is both the least understood and the most fascinating to virologists and clinicians.

The purpose of this introductory chapter is to describe the normal structure and functions of the liver and the array of alterations that can be induced by acute and chronic infections with HBV.

1 Liver Anatomy

1.1 Gross Anatomy

The liver is a large solid organ, weighing 1200–1500 g in adults, located in the right upper quadrant of the abdomen and partially protected by the ribs (SHERLOCK 1989). It is composed of a large right lobe, which is subdivided into quadrate and caudate lobes, and a smaller left lobe (Fig. 1). While the two main lobes are distinguished by ligaments, the entire organ is a contiguous parenchymal mass and the lobes are not delimited by any fibrous septum that traverses the liver substance. The vascular and biliary systems of the different lobes are interconnected (RAPPAPORT 1987).

Blood is supplied to the liver in a unique fashion that is well suited to several aspects of its functions, particularly those involving nutrient metabolism and drug detoxification. Well-oxygenated arterial blood is supplied by the hepatic artery, while venous blood is carried to the liver from the intestines and spleen by the portal vein. The portal system brings absorbed carbohydrates, fats, and proteins, as well as drugs and other toxins directly from the gut to the liver. Both the portal vein and hepatic artery enter the liver at the porta hepatis where they divide to form separate branches to the left and right hepatic lobes. Blood is carried from the liver by two or more hepatic veins that open into the inferior vena

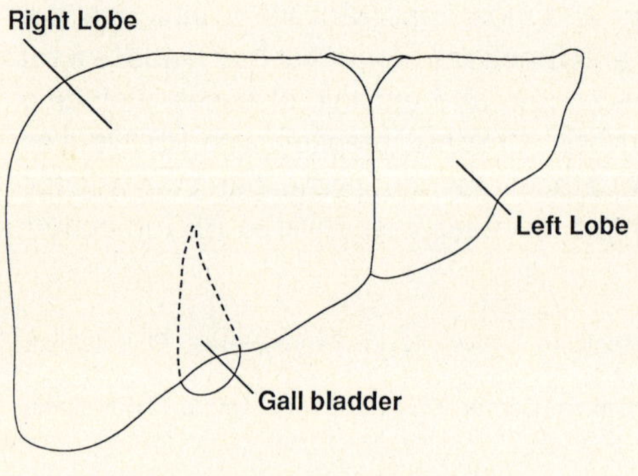

Fig. 1. Anterior view of the human liver. The liver fits snugly under the dome of the diaphragm and is protected anteriorly by the rib cage. With the ribs removed, only the right and left lobes of the liver and the inferior pole of the gall bladder are visible. The quadrate and caudate lobes are subdivisions of the right lobe located on the posterior (caudate) or inferior (quadrate) aspects of the liver

cava near where it enters the heart. Lymph vessels from the liver terminate in glands near the porta hepatis, along the inferior vena cava in the thorax and mediastinum (SHERLOCK 1989).

The biliary system, which originates in the liver, supplies bile to the small intestine where it aids in fat absorption. Bile is secreted by hepatocytes into the bile canaliculi, transported and modified by the intrahepatic ductules to exit the liver through the common hepatic duct (ERLINGER 1987). From the common hepatic duct, bile flows into the common bile duct. The cystic duct is a branch off the common bile duct that connects it with the gall bladder. The gall bladder, while not technically part of the liver, is anatomically and functionally related to it. Located on the under surface of the right lobe of the liver, the gall bladder is a pear-shaped organ that has a capacity to hold 50 ml of fluid. It stores hepatic bile until a hormonal signal, in response to food in the small intestine, induces contraction, causing bile to course back through the cystic duct into the common bile duct. From there, bile continues distally and empties into the intestine (duodenum) through the ampulla of Vater, a channel formed by the confluence of the common bile duct and the pancreatic duct.

Shortly after food enters the small intestine, bile and pancreatic juice are secreted into it. These secretions act on the nutrients to begin the process of digestion. The major constituents of bile are bile acids which are synthesized in hepatocytes and secreted into bile ductules. After aiding in the digestive process, which occurs throughout much of the small intestine, the bile acids are actively reabsorbed in the terminal ileum (the very end of the small intestine). After absorption, the bile acids enter the portal system and are efficiently removed by the liver where they are promptly secreted into bile once again. The entire anatomic route taken by these important transport molecules is called the enterohepatic circulation (HOFFMAN 1989).

1.2 Microscopic Anatomy

The liver has been divided into structural subunits, called lobules, based primarily on its two-dimensional appearance in the light microscope. First described by KIERNAN in 1833, a lobule consists of a polyhedral section of liver tissue which appears hexagonal on cross-section. It is composed of a central vein (a tributary of the hepatic vein) and, at the periphery, portal tracts (triads) each of which contains a bile duct, portal vein tributary, and a branch of the hepatic artery (Fig. 2). The central hepatic vein and the portal triads are connected by cords of liver cells and sinusoids (contiguous spaces between the liver cells). In three dimensions, the cords of liver cells are really sheets of hepatocytes, one cell in thickness (ELIAS 1949; GRISHAM 1976). Under normal circumstances the portal vein has a higher pressure than the central venous system so that blood flows from the portal system into and through the sinusoids towards the central hepatic vein (SHERLOCK 1989). The arterial system follows the same path as the portal system resulting in a mixing of arterial and portal venous

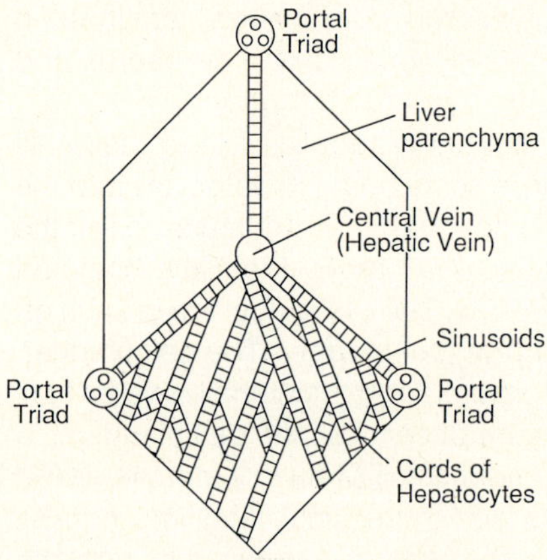

Fig. 2. Schematic representation of a hepatic lobule based on two-dimensional light microscopy. The central vein in the center of the hexagonal lobule is a tributary of the hepatic venous system. Each portal triad contains a tributary of the portal vein, a branch of the hepatic artery, and a bile duct. The sinusoids are blood-filled spaces between cords of hepatocytes. Both portal and arterial blood flows into the sinusoids and from there into the central veins (SHERLOCK 1989)

blood in the sinusoids. Sphincters in each system regulate flow into individual sinusoids.

Since their original description, the in vivo existence of hepatic lobules has been questioned (AREY 1932; BRITTAIN et al. 1964; OPIE 1944; MALL 1906; RAPPAPORT 1953; RAPPAPORT et al. 1954). The concept of a microscopic unit of liver tissue filling a hexagonal space around a central vein breaks down when one attempts to integrate these individual units into an overall structure for the liver. In addition, the concept of a hexagonal lobule is not compatible with observations of the hepatic microcirculation in situ (RAPPAPORT 1987).

Based on studies of the microcirculation, the liver has been divided into microscopic units called acini (RAPPAPORT 1976). Each acinus is an irregular mass of parenchymal tissue centered on a portal triad which again consists of a portal vein, bile duct, and hepatic artery. The acinus lies between and drains into two hepatic (central) veins (Fig. 3). It has no surrounding capsule and by two-dimensional light microscopy appears to occupy part of two adjacent hepatic lobules. The liver parenchyma immediately surrounding the portal triad is referred to as zone 1. Zone 3 lies at the periphery of the acinus, near the hepatic veins. This zone receives the poorest supply of oxygenated blood and is the area that suffers the most injury from anoxic, toxic, or viral insults (SHERLOCK 1989). Zone 2 is located between zones 1 and 3 and is intermediate, as well, in terms of blood supply.

In addition to hepatocytes, the liver is composed of endothelial cells, Kupffer cells, and bile duct epithelial cells. Sixty percent of the cells in the liver are hepatocytes. These are polygonal cells approximately 30 µm in diameter. In relation to their environment, hepatocytes have three surfaces, but no basement membrane. One surface of the hepatocyte faces the sinusoid and the space of Disse, while a second surface borders the bile cannaliculus, and the third touches neighboring liver cells. The plasma membrane of hepatocytes is reinforced by microfilaments forming a supportive cytoskeleton. The surface of hepatic plates is sealed by junctional complexes between individual hepatocytes, including tight

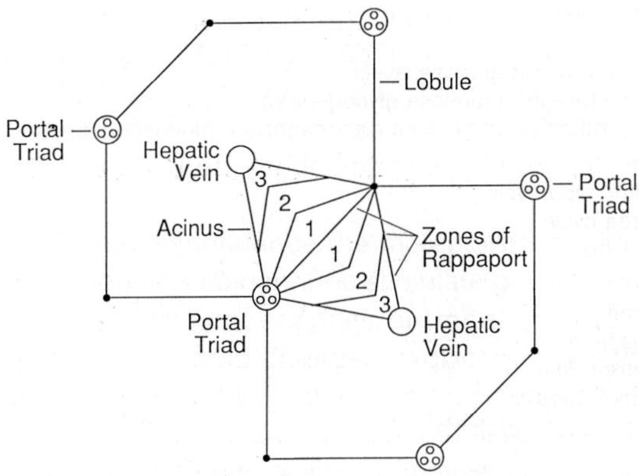

Fig. 3. Schematic representation of the hepatic acinus as proposed by RAPPAPORT (1976). The hepatic acinus is centered on a portal triad and lies between two hepatic (central) veins. The parenchymal tissue closest to the portal triads was designated by Rappaport as zone 1. It receives the best-oxygenated blood and is the area most resistant to liver injury. Zone 3 is closest to the hepatic veins, receives the poorest supply of oxygenated blood, and is most susceptible to liver injury (RAPPAPORT 1976; SHERLOCK 1989). Zone 2 is intermediate in oxygen supply and intermediate in susceptibility to liver injury

junctions, gap junctions, and desmosomes. In a healthy liver an occasional mitosis can be seen. The lifespan of a liver cell in experimental animals is about 150 days.

The sinusoids are hepatic capillaries lined with endothelial cells. Kupffer cells are phagocytic cells that form part of the sinusoids and comprise a major part of the reticuloendothelial system. The space of Disse is a tissue space between the hepatocytes and the sinusoidal lining cells. Bile canaliculi form as grooves on the contact surfaces of liver cells. The canalicular network drains near the portal vein into thin-walled bile ductules that are lined with cuboidal epithelial cells. The ductules terminate in the bile ducts (SHERLOCK 1989).

2 Major Hepatic Functions

Normal liver function is essential for human life. While technology has made it possible for individuals without kidneys to continue living through hemo- or peritoneal dialysis, no such techniques exist to compensate for the loss of normal hepatic function. The liver is essential for carbohydrate, protein, and lipid metabolism. Furthermore, it plays an important part in bilirubin, drug and toxin, vitamin and metal metabolism (Table 1). Severe liver disease, such as cirrhosis from a variety of causes, results in predictable symptoms from derangements in these metabolic pathways. While no artificial means exists to assume the essential metabolic activities of the liver, orthotopic liver transplantation in selected

the proteins synthesized by the liver are the acute phase response factors: albumin, fibrinogen, alpha-1-antitrypsin, heptaglobin, ceruloplasmin, and transferrin. They rise in response to tissue injury, which may contribute to maintaining serum concentrations of these proteins even in the face of liver disease.

Approximately 10 g albumin are synthesized by the liver each day. Albumin has a 22 d half life, and its serum concentration falls only gradually after onset of liver disease, even if quite severe. Alpha-1-antitrypsin is a genetically polymorphic trypsin inhibitor; the inheritance of a particular form of this protein results in a deficiency that is associated with liver and pulmonary disease. Haptoglobin is a glycoprotein and is lowered in hemolytic diseases and hepatocellular failure. Ceruloplasmin is the major copper-containing protein in the blood, and a low concentration is associated with Wilson's disease, a liver and brain disease that results from copper overload of those organs. Transferrin is the major iron transport protein. It is greater than 90% saturated in patients with hemochromatosis, a disease of iron overload in multiple organs, including the liver. Alpha-fetoprotein is a normal component of plasma in human fetuses and reaches a maximum concentration at about 16 weeks of fetal life. Shortly after birth it disappears from the blood. It is synthesized again by many hepatocellular carcinomas and is used clinically for the diagnosis and early detection of such tumors.

The liver is the principal site of synthesis of all coagulation factors except the von Willebrand factor and the fibrolytic factors (FAIR and MARLAV 1986; KELLY and SUMMERFIELD 1987). The liver also synthesizes protease inhibitors that modulate the coagulation cascade, such as antithrombin III, protein C, and heparin cofactor II. The liver uses vitamin K in the synthesis of many of the important blood coagulation factors. The half-life of these clotting proteins is short, and therefore they become depleted rapidly after the onset of hepatic dysfunction. Accordingly, other factors, such as thrombocytopenia, disseminated intravascular coagulation, and failure of bile salt secretion into the intestine resulting in inadequate absorption of vitamin K can all contribute to abnormal bleeding in patients with severe liver disease. Clinically, this is often manifested by bleeding from the gastrointestinal tract.

Different hormones exert a variety of different effects on the liver. Most hormones have a specific receptor on the cell surface or in the cytoplasm. Liver response to a given hormone depends on the concentration of the hormone as well as the number and characteristics of the receptors for the hormone. Membrane-bound hormones usually act through the adenylate cyclase system and convert adenosine triphosphate (ATP) to cyclic adenosine monophosphate (AMP) which is the intracellular mediator for hormone action. Cyclic AMP works through a protein kinase producing the characteristic response of the hepatocyte. Sex hormones, on the other hand, bind to receptors in the cell cytoplasm. The hormone receptor complex then is transported to the nucleus and influences gene expression (VAN THEIL et al. 1987).

2.3 Lipids

The liver plays a role in the metabolism of the major types of lipid carried in plasma, including cholesterol, phospholipids, triglycerides, and lipoproteins. Cholesterol, synthesized in the liver from acetate in the cytoplasm of hepatocytes, is the source of virtually all of the endogenously produced sterol. It is found in cell membranes and is a precursor of bile acids and steroid hormones. The genetically induced hypercholestolemias are the result of continued cholesterol synthesis by the liver in the face of high blood levels. Occasional patients with hepatocellular carcinoma will manifest the same defect and develop very high serum cholesterol levels. In liver disease associated with bile duct obstruction, total serum cholesterol rises and sometimes is deposited in the skin (xanthomas). In many hepatocellular diseases triglycerides tend to be elevated with the excess found in the low density lipoprotein (LDL) fraction (SHERLOCK 1989).

2.4 Bilirubin Metabolism

Bilirubin is the metabolic end product of hemoglobin. After breakdown of red blood cells in macrophages of the reticuloendothelial system, the heme portion of the hemoglobin molecule is catabolysed to biliverdin which is rapidly converted to bilirubin and transported to the liver. In the liver, bilirubin is conjugated to glucuronic acid or tightly bound to albumin and is then secreted into the bile. Bilirubin diglucuronide is the major conjugate in human bile. The secretion of bile, by hepatocytes, into the cannalicular system is controlled by at least two independent processes. One of these mechanisms is dependent on bile salts, while the other depends on organic anions such as bilirubin. Bilirubin glucuronide is polar and therefore not absorbed from the small intestine. In the colon it is broken down to urobilinogen which is nonpolar and readily absorbed. Reabsorbed urobilinogen is transported by the portal circulation to the liver, where it is taken up by hepatocytes and re-excreted into the biliary tree or into the plasma where it is carried in the general circulation to the kidney and excreted in urine. With advanced liver disease, re-excretion by the liver is impaired, and more urobilinogen ends up in the urine turning it dark green. Impairment of the re-excretion of urobilinogen is also responsible for symptoms of fever and heart failure in the early stages of viral hepatitis. Hepatocellular disease may result in jaundice by: disturbing the uptake and transport of bilirubin in the hepatocyte, causing a defect in conjugation, inducing abnormalities in the cell membrane opposite the canaliculus for excretion of bile, or by obstructing the large bile channels before they reach the intestine. Once an excess amount of bilirubin is in the circulation, clinical jaundice, including symptoms of dark skin and pruritis, results when bilirubin enters tissues. Circulating protein-bound bilirubin does not readily enter tissue fluids that are low in protein. Therefore, exudates tend to be more icteric than transudates. Skin, ocular sclera, and blood vessels have a high elastin content and easily become jaundiced (SHERLOCK 1989).

2.5 Drug and Toxin Metabolism

The liver is the principal organ involved in drug and toxin metabolism. Oral drugs that are lipid soluble pass the membrane of the intestine and are delivered to the liver via the portal system. In the liver they are converted to water-soluble compounds for excretion via urine or bile. Hepatic drug metabolism to such products can be divided into two phases. Phase 1 is the main drug-metabolizing system and resides in the smooth endoplasmic reticulum of the liver cell. The enzymes involved include the mixed function mono-oxygenases, cytochrome C reductase, and cytochrome P450. Reduced nicotinamide-adenine-dinucleotide phosphate (NADPH) is a cofactor. The drug is oxidated or hydoxylated to a more polar compound. Alternative phase 1 drug-metabolizing reactions include the conversion of alcohol to acetaldehyde by alcohol dehydrogenases. Phase 1 enzymes can be induced or inhibited by a variety of compounds.

Phase 2 metabolism of drugs involves polarization via conjugation with small endogenous molecules (e.g., glucuronide, sulfate). Although the enzymes catalyzing phase 2 reactions are found elsewhere in the body, they are usually most abundant in the liver. In patients with liver disease, the half-lives of drugs are often prolonged in correlation with the extent of hepatocellular damage (FARRELL et al. 1978). In these patients, side effects associated with excessive retention of certain hepatically cleared drugs may occur at much reduced dosages. Other drugs, which can cause liver toxicity by forming toxic intermediates, are also more likely to do so at a reduced dose in patients with already impaired liver function (SHERLOCK 1989).

3 Pathology

3.1 Patterns of Hepatic Injury

Hepatocytes are subject to injury by three distinct mechanisms: direct toxic effects of chemicals or metabolites; indirect, immunologically mediated damage; and the effects of bile retention (cholestasis) (SHERLOCK 1985). In humans with liver disease, all three mechanisms frequently operate simultaneously. This results, at the histopathologic level, in a mixture of lesions which makes assigning causes of liver injury by morphologic criteria alone virtually impossible.

Furthermore, hepatocytes have only a few ways of responding to injury. These are usually described in the light microscope as ballooning degeneration, eosinophilic (acidophilic) degeneration, hyaline degeneration, and fatty degeneration or steatosis. Ballooning is characterized by swelling of liver cells, clear or finely granular cytoplasm, and enlarged, irregular, pale-staining nuclei. In the electron microscope, the cisternae of the endoplasmic reticulum are dilated, the cytoplasm is degranulated, and the glycogen content is reduced. Balloon cells are thought to have a very limited survival, rapidly undergoing lytic necrosis, breakdown of cell membranes, and disappearance. Ballooning is commonly seen in association with immune-mediated damage and cholestasis.

In eosinophilic degeneration, the affected hepatocytes are shrunken rather than swollen; the cytoplasm is dark pink (eosinophilic) with hematoxylin and eosin (H and E) stains; and the nuclei are shrunken, darkly stained, and structureless (pyknotic). These cells continue to shrink, lose their nuclei, and are extruded from liver cell plates into sinusoidal spaces. There they may persist as deeply eosinophilic-staining bodies of varying sizes and shapes. These structures were first described by Councilman (and are often called "Councilman bodies") (ISHAK 1976).

Hyaline degeneration and steatosis are best understood in the context of direct injury to the liver. In this process, the damage is primarily to cytoplasmic organelles including mitochondria, endoplasmic reticulum, and microtubules. Alcohol is the classic example of a directly toxic agent. Metabolites of alcohol (e.g., acetaldehyde) damage the mitochondrial membrane. In the electron microscope the mitochondria are enlarged and distorted. Hyaline degeneration is frequently indicated in pathologist's reports by the presence of Mallory bodies (also called alcoholic hyaline) (MALLORY 1911). These bodies appear in the light microscope as eosinophilic masses of hyaline which may occupy most of the cytoplasm. Ultrastructurally, Mallory bodies are composed of filaments of varying sizes and shapes. These, in turn, are probably related to abnormalities in the synthesis and function of microtubules (SHERLOCK 1985).

Steatosis, the accumulation of fat in hepatocytes, is often prominent in alcoholic liver disease. There are several metabolic abnormalities that contribute to the accumulation of fat including increased synthesis of triglycerides and lipoproteins, lipid peroxidation, and decreased activity of the citric acid cycle (NILIUS 1985).

These changes in hepatocytes are accompanied by a cellular infiltrate of polymorphonuclear leukocytes and proliferation of Kupffer cells. Taken together, these abnormalities are representative of the spectrum of effects of direct liver injury. None of these abnormalities is specific for alcoholic liver disease. Any or all of them may occur in other, widely disparate diseases such as diabetes mellitus, obesity, and Wilson's disease (a disorder of copper metabolism) (SHERLOCK 1985).

Viral hepatitis demonstrates the indirect, immune mechanism of damage to hepatocytes. The injury is to the cell membrane and is due to specific and nonspecific cytotoxic effects of lymphocytes. With the light microscope, the most prominent abnormality is a rich infiltrate of lymphocytes and plasma cells in portal areas and within lobules. Liver cells show evidence of balloon degeneration, eosinophilic degeneration, and necrosis as well as regeneration (BIANCHI et al. 1979). The stimulus to regeneration is cell death. Thus, hepatocyte necrosis is always accompanied by increased numbers of cells in mitosis, the hallmark of regeneration. In the normal mature liver, only one cell in 10000 is in mitosis (BECKER 1974).

Cholestatic injury is characterized by arrested bile and, at the light microscopic level, minimal hepatocyte damage, mainly accumulation of bile in liver cells and in canaliculi. Ultrastructurally, however, the lesions of cholestasis are not trivial. The Golgi apparatus is vacuolated, lysosomes are increased in number, and the endoplasmic reticulum is hypertrophied. One hypothesis is that these changes are brought about by the retention of bile salts. Cholestasis can occur as the result of either mechanical obstruction to the outflow of bile (e.g., a stone in the common bile duct) or as a functional impairment of the secretion of bile by hepatocytes. Cholestasis may occur as a response to drugs or chemical exposures, or as part of the spectrum of virus-induced lesions (SHERLOCK 1985).

4 Pathology of Liver Diseases Associated with Hepatitis B

4.1 Acute Hepatitis

The term "acute hepatitis" refers to a disease characterized by diffuse inflammation of the liver. It is not synonymous with "transient" hepatitis. Hepatitis viruses A,B,C, and E can all cause acute hepatitis. The disease is always transient when caused by the A or E viruses, but can be either transient or persistent when induced by the B or C viruses. Even though only acute hepatitis B or C may progress to chronic forms of hepatitis, the histopathology of acute viral hepatitis is indistinguishable whether caused by hepatitis viruses A, B, C, or E. The entire liver is involved, usually to the same degree in all areas. Four morphological patterns may occur that refer to the location and extent of hepatocellular necrosis and are, hence, approximately related to the severity of the disease. They are: classical hepatitis with spotty necrosis, hepatitis with bridging necrosis, with panlobular necrosis, or with periportal necrosis (SCHEUER 1985).

In classical acute hepatitis, hepatic damage is focal. Liver cell necrosis involves single hepatocytes or small groups of cells, particularly in the centers of lobules. Balloon degeneration, eosinophilic degeneration, and Councilman bodies can be detected. Portal tracts and sinusoids are filled initially with lymphocytes and later, as the lesion evolves, with a mixed population of inflammatory cells including lymphocytes, polymorphonuclear granulocytes, and monocytes. The phagocytic cells scavenge the necrotic debris, but they may also contribute to the damage to hepatocytes by releasing the products of oxidative metabolism (BIANCHI et al. 1979; SCHEUER 1985).

Degenerative and regenerative changes are both present in lobules. Liver cells vary in size and shape (pleomorphism) with several mitoses and binucleated cells seen in each high-powered field. There is diffuse enlargement of sinusoidal lining cells and focal areas of hyperplasia. It is thought that the lining cells are released into the sinusoids and become macrophages.

Bile pigment is not always present. When it is, it can occur as bile thrombi in canaliculi, as fine or coarse granules in hepatocytes, or within Kupffer cells.

As the acute stage of hepatitis resolves, the liver shows a restoration of normal parenchymal architecture. Some areas of spotty necrosis may persist, along with inflammatory cells in the portal areas and around hepatic veins in lobules. Fine reticular fibers between portal and central zones and between portal tracts may be seen at this stage, but they do not progress to cirrhosis (BIANCHI et al. 1979).

In acute hepatitis with bridging necrosis, the extent of hepatocyte damage is greater than in classical acute hepatitis. There are bands of confluent necrosis between centrilobular areas and portal tracts (BOYER and KLATSKIN 1970). According to RAPPAPORT'S (1976) description of the microanatomy of the hepatic acinus, zone 3 receives the poorest blood supply from the terminal branches of the portal vein and the hepatic artery. It is this zone of relatively poor blood supply that is most affected by bridging necrosis. Balloon degeneration with rapid lysis is the major hepatocyte lesion, which results in the "dropping out" of hepatocytes from hepatic plates. The bands of necrosis may vary in width, cellularity, and presence or absence of septae. The cellular infiltrate, which is predominantly composed of phagocytic cells (polymorphonunclear leukocytes and monocytes) rather than lymphocytes, may reflect the speed of necrosis rather than the mechanism of injury.

Clinically, the significance of bridging necrosis is that it carries a much greater risk of a poor outcome than classical hepatitis with spotty necrosis. Although in some cases it may heal and leave a normal liver, more frequently, in the context of HBV or HCV infection, it may result in chronic hepatitis, cirrhosis, or death from hepatic failure (BOYER and KLATSKIN 1970).

Acute hepatitis with panlobular (massive) necrosis is an even more severe lesion. The necrosis involves entire acini or entire lobules. When the whole liver is affected, it is recognized clinically as fulminant hepatitis and carries a 70% risk of death. Survival apparently depends on rapid liver cell regeneration and is reflected clinically by a rise in the serum alpha-fetoprotein concentration (KARVOUNTZIS and REDEKER 1974). Fortunately, panlobular necrosis is relatively rare; less than 1% of cases of acute hepatitis become fulminant.

Periportal necrosis is characterized by extensive cellular infiltration and necrosis in periportal and portal areas out of proportion to other areas. There may be portal to portal bands of necrosis. POPPER et al. (1965) described necrosis of liver cells at the interface of parenchymal and connective tissue and gave it the name "piecemeal necrosis." This type of necrosis is associated with, and may give rise to, periportal necrosis. Patients with periportal necrosis infected with HBV or HCV are at high risk of developing chronic hepatitis and cirrhosis (BIANCHI et al. 1979).

Hepatitis B viral antigens, when present, can be detected in either frozen sections or formalin-fixed, paraffin-embedded liver tissues by immunofluorescence or immunoperoxidase methods. In classical acute hepatitis B, it is frequently difficult to detect viral markers in tissues, particularly at the height of lobules are infiltrated with lymphocytes, necrosis is focal and mild, and portal infiltration is not prominent. Progression to CAH or cirrhosis is unusual. The distribution of viral antigens is similar to that seen in CPH.

4.3 Cirrhosis

Cirrhosis is a type of host response to chronic liver injury that is characterized by bands of fibrosis throughout the liver associated with and surrounding nodules of parenchymal tissue. The initiating event is necrosis of hepatocytes. In "active" cirrhosis there is continuing liver cell damage and death which provides the stimulus for hepatocyte regeneration, nodule formation, and laying down of more fibrous tissue. "Inactive" cirrhosis indicates that hepatocyte damage has ceased or has been reduced to a low level, resulting in arrest of the other components of the disease process (SCHEUER 1979). As with acute and chronic hepatitis, cirrhosis is not a specific manifestation of chronic viral infection. It can be caused by drugs, alcohol, iron (hemochromatosis), retention of bile constituents (biliary cirrhosis), as well as chronic hepatitis B or C.

Piecemeal necrosis is commonly seen in active cirrhosis of whatever cause, as well as cholestasis, steatosis, and iron accumulation. Nodule formation is the result of over-activity of the regeneration process. That is, the stimulus to cell division is cell death. If the replacement process were precise, there would be a one-to-one exchange of dead cells for new cells. In cirrhosis, however, cell division usually continues beyond replacement giving rise to small (less than 3 cm in diameter) or large hyperplastic nodules (micro- or macronodular cirrhosis), which lack the normal association into hepatic plates (ANTHONY et al. 1978). HBV and HCV are usually associated with macronodular and alcohol abuse with micronodular cirrhosis.

Cirrhosis in its early stages is usually asymptomatic, but with time patients develop effects that are often fatal. As nodules form and grow, they distort the architecture of lobules causing compression of hepatic venules. Fibrosis in portal areas may obstruct portal venous blood flow. Both fibrosis and nodule formation increase resistance to blood flow through sinusoids. Shunts may develop between portal venules and hepatic arterioles. These changes increase the pressure in the portal vein and its tributaries (portal hypertension) which in turn causes the spleen to enlarge (splenomegaly) and to induce dilatation of small veins at the lower end of the esophagus and upper end of the stomach (esophageal varices) (SCHEUER 1979). These veins are thin walled and are prone to rupture causing massive hemorrhage. Rupture of esophageal varices is a frequent cause of death of patients with cirrhosis.

A second cause of morbidity and mortality in cirrhosis is dysfunction of hepatocytes. The various vascular changes in the liver reduce access of liver cells to nutrients and oxygen. Continuing hepatocyte necrosis eventually reduces the number of functioning hepatocytes. These abnormalities result in a decreased synthesis of albumin which, in turn, decreases the intravascular osmotic pressure. Portal hypertension and low osmotic pressure result in fluid losses from mesenteric capillaries and leakage of hepatic lymph into the peritoneal cavity (ascites). Failure of hepatocytes to conjugate and excrete bile leads to an accumulation of bilirubin in the circulation and in the skin (jaundice). The failure of the liver to clear various metabolites, particularly ammonia, from the circulation may result in disorientation, coma, and death (MACSWEEN and SCOTT 1973).

HBV antigens have an interesting distribution in cirrhotic livers. Each nodule appears to have its own pattern. Some nodules are free of viral markers, while in others the cytoplasm of every cell stains intensely and homogeneously for surface antigen. Some nodules show a uniformly membranous distribution of HBsAg with little cytoplasmic staining. HBcAg is less commonly demonstrated and, when seen, is usually in nuclei (NAYAK et al. 1977; NAZAREWICZ et al. 1977). In some nodules however, all cells may have a cytoplasmic distribution of core antigen (ZHOU et al. 1987). This heterogeneity of expression may reflect different degrees of maturation of hepatocytes among regenerating nodules.

5 Hepatocellular Carcinoma

Patients with cirrhosis of any etiology, but particularly those chronically infected with HBV or HCV, are at increased risk of developing a cancer in the liver. These tumors are often called "primary" hepatocellular carcinomas (PHC) to clearly distinguish them from tumors that arise in other organs and metastasize to the liver.

According to the World Health Organization, about 80% of the cases of PHC in the world are associated with chronic HBV infection (WHO 1983). Also, about 80% of PHCs arise in a setting of cirrhosis. In another 15% of cases the liver is affected by chronic hepatitis (CAH or CPH). Less than 5% of cases occur in a histologically normal liver (LONDON 1981).

In the past it was thought that most PHCs were multicentric (i.e., multiple primary tumors). Recent studies of patients with PHCs, diagnosed at an early stage, indicate that most or all of these tumors begin as solitary nodules (TANG 1985). The tumors consist of parenchymal cells which may be well, moderately, or poorly differentiated. The tumor cells are often arranged in trabeculae two to eight cells thick, separated by sinusoids. Kupffer cells are usually absent from tumor nodules (PETERS 1976).

PHCs are highly lethal, usually causing the death of the patient within 6 months of diagnosis. The tumors commonly invade the portal and venous systems allowing rapid spread throughout the liver. Death is generally the result of hepatic failure rather than from metastases to other vital organs.

Viral antigens are distributed differently in the cancerous and noncancerous parts of livers containing PHC. Because PHCs are usually associated with

Mallory FB (1911) Cirrhosis of the liver, five different lesions from which it may arise. Bull Johns Hopkins Hosp 22: 69–75
Mondelli M, Vergani GM, Alberti A, Vergani D, Portmann B, Eddelston ALWF (1982) Specificity of T-lymphocyte cytotoxicity to autologous hepatocytes in chronic hepatitis B infection. Evidence that T cells are directed against HBV core antigen expressed on hepatocytes. J Immunol 129: 2773–2778
Nayak NC, Dhan A, Sachdeva R, Mittal A, Seth HN, Sudarsanam D, Reddy B, Waghaolikar UL, Reddy CRRM (1977) Association of human hepatocellular carcinoma and cirrhosis with hepatitis virus surface and core antigens in the liver. Int J Cancer 20: 643–654
Nazarewicz T, Krowczynski K, Slusarczyk J, Nowoslawski A (1977) Cellular localization of hepatitis B virus antigen in patients with hepatocellular carcinoma coexisting with liver cirrhosis. J Infect Dis 153: 298–302
Nilius R (1985) Acetaldehyde, aldehyde dehydrogenase, and pathogenetic aspects of alcoholic liver disease. In: Brumer H, Thaler H, Festschrift for Hans Popper. Raven, New York
Opie EL (1944) The pathogenesis of liver tumors produced by butter yellow. J Exp Med 80: 231–246
Peters RL (1976) Pathology of hepatocellular carcinoma In: Okuda K, Peters RL (eds) Hepatocellular carcinoma, ed. Wiley, New York
Popper H, Paranetto F, Schaffner F (1965) Immune processes in the pathogenesis of liver disease. Ann NY Acad Sci 124: 781–799
Rappaport AM (1953) In discussion of Elias H: morphology of the liver, Trans 11th Liver Injury Conf, Macy, New York
Rappaport AM (1976) The microcirculatory acinar concept of normal and pathological hepatic structure. Beitr Pathol 157: 215
Rappaport AM (1987) Physioanatomic considerations. In: Schiff L, Schiff E (eds) Diseases of the liver. Lippincott, Philadelphia
Rappaport AM, Borowy ZJ, Lougheed WM, Lotto WN (1954) Subdivision of hexagonal liver lobules into a structural and functional unit: role in hepatic physiology and pathology. Anat Rec 119: 11–27
Scheuer PJ (1979) Chronic hepatitis. In: MacSween RNM, Anthony PP, Scheuer PJ (eds) Pathology of the liver: Churchill Livingstone, Edinburgh London New York
Scheuer PJ (1985) Pathological aspects of viral hepatitis. In: Brummer H, Thaler H (eds) Hepatology, Festschrift for Hans Popper. Raven, New York
Sherlock S (1985) Patterns of hepatic injury in man. In: Brunner H, Thaler H (eds) Hepatology, Festschrift for Hans Popper. Raven, New York
Sherlock S (1989) Anatomy and function. In: Sherlock S (ed) Diseases of the liver and biliary system. Blackwell Scientific, Cambridge
Tanikawa K (1979) Virus particles. In: Ultrastructural aspects of the liver and its disorders, 2nd edn. Igakushoin, Tokyo
Tang ZY (1985) Subclinical hepatocellular carcinoma-historical aspects and general considerations. In: Tang ZY (ed) Subclinical hepatocellular carcinoma. Chinese Academic, Beijing, Springer-Verlag, Berlin
Van Theil D, Stone BG, Schade RR (1987) The liver and its effect on endocrine function in health and diseases. In: Schiff L, Schiff E (eds) Diseases of the liver. Lippincott, Philadelphia
WHO Scientific Group on Prevention and Control of Hepatocellular Carcinoma (1983) Prevention of primary liver cancer. Report of a meeting of a WHO scientific group. Lancet 1: 463–465
Zhou X-D, DeTolla L, Custer RP, London WT (1987) Iron, ferritin, hepatitis B surface and core antigens in the livers of Chinese patients with hepatocellular carcinoma. Cancer 59: 1430–1437

Transcriptional Control of Hepadnavirus Gene Expression

H. SCHALLER and M. FISCHER

1 Introduction

Hepadnaviruses multiply their DNA genome by reverse transcription of an RNA intermediate, termed pregenomic RNA. Therefore, hepadnaviral transcription has a dual function: the production of the RNA template for genome replication and the synthesis of messenger RNAs (mRNAs). The very small viral genome is fully covered by extensively overlapping coding sequences, requiring an expression strategy that uses differential transcription initiation and unspliced transcripts to produce at least seven viral gene products from four open reading frames. This complex organization implies that the hepatitis B virus (HBV) genome contains several transcriptional control regions, that these regions are located at short distances from each other (maximally 1.6 kb apart in the circular 3.2-kb genome), and that all are themselves actively transcribed during viral replication and eventually translated into protein. This scenario,

---

Zentrum für Molekulare Biologie, University of Heidelberg, Im Neuenheimer Feld 282, 6900 Heidelberg, F.R.G.

Current Topics in Microbiology and Immunology, Vol. 168

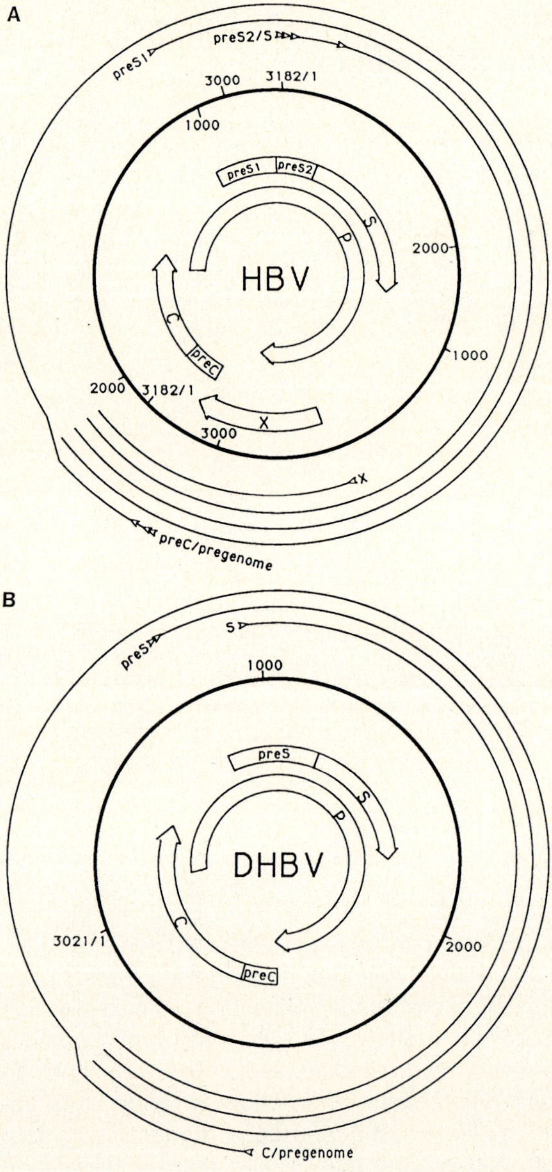

Fig. 1 A, B. Transcriptional maps of the HBV (**A**) and DHBV genomes (**B**) The RNAs are displayed as *thin lines*. *Small arrowheads* represent transcriptional start sites. Open reading frames are drawn as *open arrows*. Nucleotide numbering for HBV is according to Pasek et al. (1979) (*inside* the circle) and Galibert et al. (1979) (*outside* the circle)

which is without precedent in other known replication units, has most likely imposed unusual restrictions on the evolution of the mechanisms regulating hepadnaviral gene expression and hence on elements involved in transcriptional control.

In addition to these difficulties arising from the necessity of expressing an optimal repertoire of gene products from a small, highly compact genome, there are also major technical limitations which have restricted our knowledge of hepadnaviral transcription for a long time. First, most studies were, and still are, being performed with HBV as the prototypic hepadnavirus; however, for this virus, to date no cell culture system exists that is permissive for all steps of the replication cycle. Secondly, hepadnaviral transcription is highly tissue specific and, therefore, early results derived from transfection of cultured cells could only be believed, with some confidence, to be related to the in vivo situation after certain differentiated hepatoma cell lines had been identified which supported production of infectious virus from cloned viral DNA genomes (see chapter by C. SEEGER et al. this volume). Consequently, much of our knowledge of HBV transcription was initially built on the physical mapping of the RNA species produced by animal hepadnaviruses in the liver of their respective hosts (BÜSCHER et al. 1985; ENDERS et al. 1985; MÖRÖY et al. 1985) and on analogous results from a single chimpanzee infected by an HBV isolate of known nucleotide sequence (CATTANEO et al. 1983, 1984; WILL et al. 1987). Much additional information on the transcriptional potential of the hepadnaviruses has been gained from transfection experiments with constructs containing all, or parts of cloned viral DNA genomes. This approach, in conjunction with site directed mutagenesis, has identified numerous *cis*-acting control elements as well as the exact position of many transcription initiation sites in the viral genome, and has also helped to clarify the structure and coding capacity of individual transcripts.

As a result of these studies a basic pattern of hepadnaviral transcription has emerged, which is now generally accepted as a central part of the viral life cycle. No significant differences are known between HBV and the other two mammalian viruses studied in detail, the woodchuck hepatitis virus (WHV) and the ground squirrel hepatitis virus (GSHV), and, as outlined in Fig. 1B, a very similar general transcriptional strategy was found with duck hepatitis B virus (DHBV), the avian prototypic virus. In this article we will therefore focus primarily on the better-studied HBV system. Data from the mammalian animal virus models will be cited only if they provide additional information, and for the avian viruses only relevant differences from the mammalian viruses will be pointed out.

2 The HBV Transcripts

Extensive circumstantial evidence indicates that viral transcripts are produced from a low number of covalently closed circular viral DNA genomes utilizing the nuclear transcription machinery of the host cell to produce a set of unidirectional and overlapping transcripts. As outlined in Fig. 1A and Table 1, four classes of

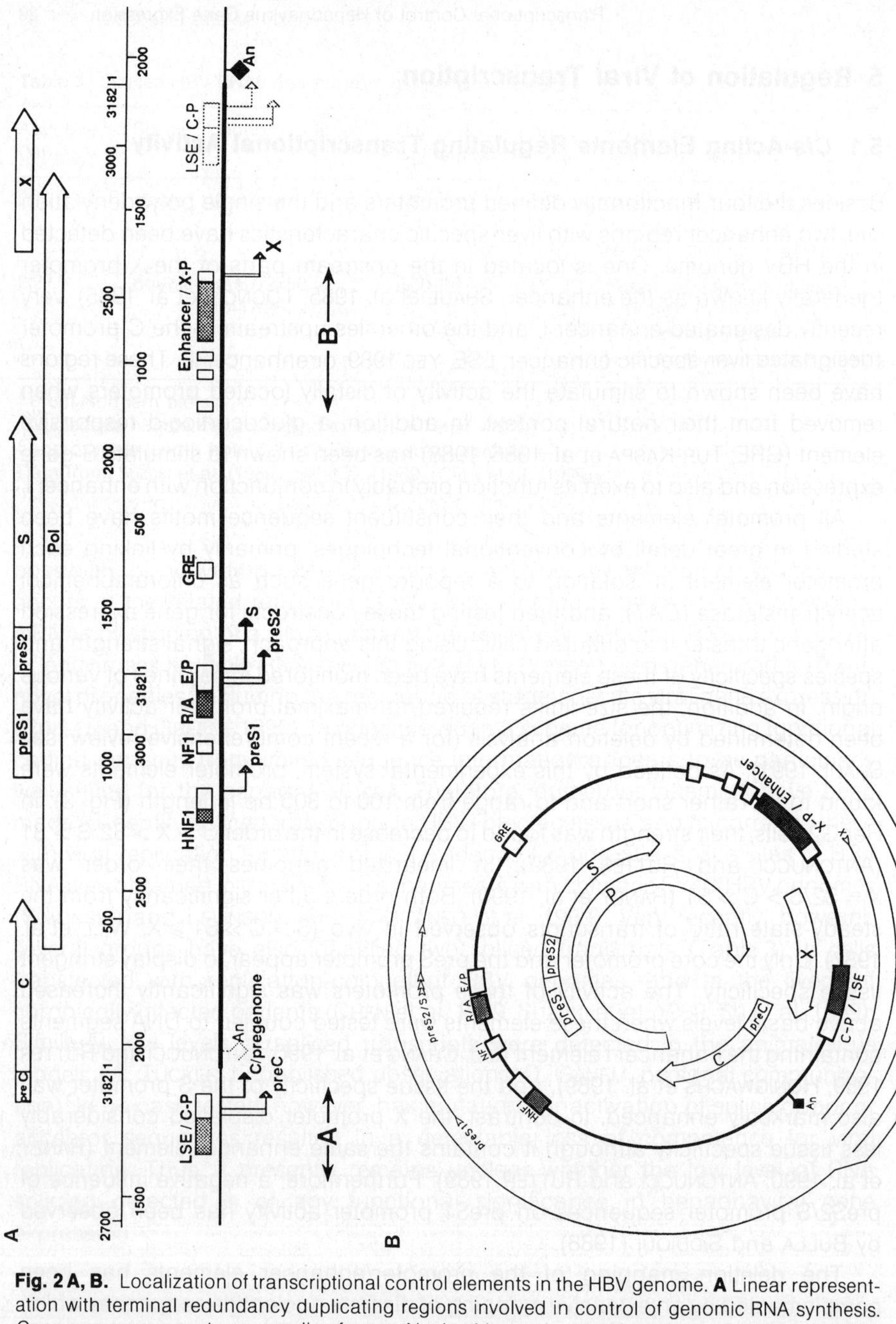

Fig. 2 A, B. Localization of transcriptional control elements in the HBV genome. **A** Linear representation with terminal redundancy duplicating regions involved in control of genomic RNA synthesis. *Open arrows* represent open reading frames. Nucleotide numbering according to Pasek et al. (1979) (*above* the line) and Galibert et al. (1979) (*below* the line). For abbreviations of control elements, see Table 4. Control elements are drawn as *boxes*, elements exhibiting liver specificity are *shaded*. Transcriptional start sites are indicated by *small arrows*. A_n, polyadenylation signal. Control elements and signals are shown in **boldface** when functionally active in the displayed position. *Lighter print* indicates inactive duplications. Protein binding sites in the areas marked *A* and *B* are shown in greater detail in Fig. 3 *A, B*. **B** Circular representation facilitating comparison with Fig.1

Table 4. Transcriptional control elements in the HBV genome

Site	Position	Comment
PreS1 Promoter region		
HNF1	2717–2739	HNF1 binding site. Also identified as AFP1 binding site [1, 2]
PreS2/S Promoter region		
NF1	2967–2986	NF1 binding site [3]
R/A	3079–3134	Sequence activates preS/S promoter in liver cells, but represses in HeLa cells (4)
E/P	3155–30	Enhancer/promoter-like element, nonspecific, suppressed by adjacent negative regulator R/A. Homologies to API, CRE, GCN4 binding sites and to SRE (LDL sterol response element) [4]
GRE	341–370	Glucocorticoid receptor binding site, biologically functional in conjunction with HBV enhancer [5]
Enhancer/X Promoter Region (Enh/X-P)		
UE3	882–901	Footprint of an unidentified protein [6]
FP VIII	977–994	Binding site of the liver-specific C/EBP [7], probably identical to UE2 (973–992) [6]
FP VII	998–1004	NFkB binding site [7]
FP VI	1026–1050	Binding site of the liver-specific C/EBP and HNF1 [8]. Coincides with UE1 (1030–1049) [6], also binds OCT2 [7]
FP IV	1101–1119	NF1 binding site, also mapped as NF1c by Ben-Levy et al. [9] to positions 1097–1114
FP V	1128–1139	Binding site of a liver-specific protein factor [8]
FP III	1153–1172	EFC binding site [8]. Also identified as EP (1145–1170) by Shaul et al. [6]
eH-TF		Binding site of eH-TF [7], also identified as TGT3 element by Ben–Levy et al. [9], overlaps FP III
NF1	1159–1189	NF1 binding site (NF1b in Ben–Levy et al. [9], overlaps FP II and III
FP II	1174–1199	C/EBP [8] and AP1 [10] binding site, cAMP induction may be mediated by binding of CREB to a CRE homology in this element [10], coincides with the element E mapped by Shaul et al. [6]
FP 1	1205–1235	NF1 binding site [8]
C-Promoter Region (C-P/LSE)		
FP I	1599–1615	Footprints induced by rat liver nuclear extracts [11]
FP II	1619–1632	
FP III	1637–1703	Footprints containing two binding sites of the liver-specific factor EBP binding sites [11]
FP IV	1708–1743	Liver-specific footprint [11], also identified as 3.6 PBP (1715–1731) by Yaginuma et al. [12], possibly a binding site of eH-TF [7]
FP V	1749–1777	Liver-specific footprints containing several EBP binding sites [11]
FP VI	1801–1827	
FP VII	1833–1850	
LSE	1685–1773	Liver-specific element (enhancer II) [13]

[a] Nucleotide numbering according to Galibert et al. (1979).

Data from: [1] Nakao et al. (1989); [2] Courtois et al. (1988); [3] Shaul et al. (1986); [4] De Medinal et al. (1988); [5] Tur Kaspa et al. (1986, 1988); [6] Shaul and Ben–Levy (1987); [7] A. Siddiqui, personal communications; [8] Patel et al. (1989); [9] Ben-Levy et al. (1989); [10] Faktor et al. (1990); [11] Lopez-Cabrera et al. (1990); [12] Yaginuma and Koike et al. (1989); [13] Yee (1989); for review see also Shaul (1991)

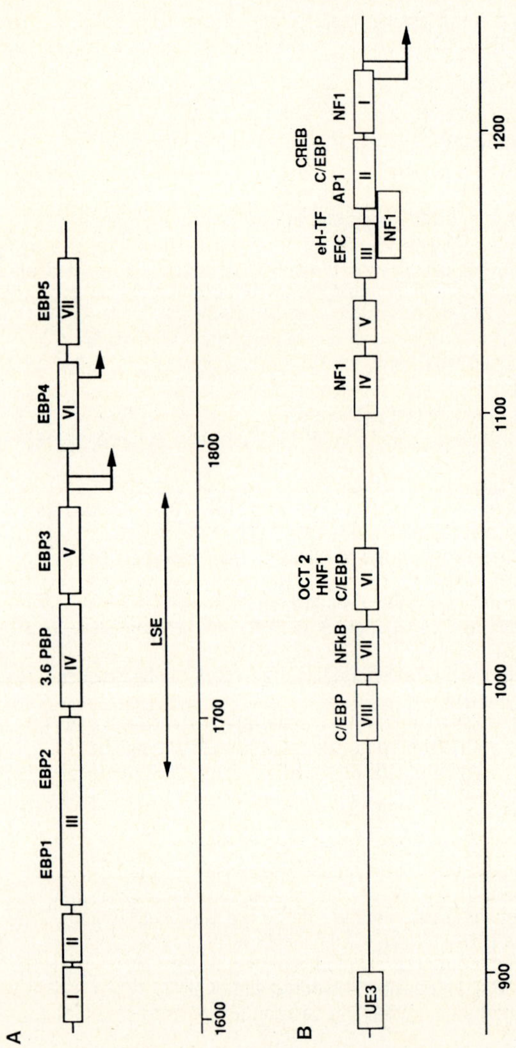

Fig. 3 A, B. Protein binding sites in the two main regulatory regions of the HBV genome. **A** C promoter/LSE. Footprints are numbered *I–VII* according to Lopez–Cabrera et al. (1990). For abbreviations, see Table 4. Nucleotide numbering according to Galibert et al. (1979). **B** Enhancer/X promoter. Footprints are numbered *I–VII* according to Patel et al. (1989). For abbreviations, see Table 4. Nucleotide numbering according to Galibert et al. (1979)

DNA sequences (NF1, C/EBP, HNF1, AP1) are structurally and functionally well defined and have also become available in pure form by recombinant DNA methods. The majority of factors, however, are defined primarily by the nucleotide sequences which they recognize (see also SHAUL 1991).

Altogether, these data reveal a dense clustering of protein binding sites along the promoter regions analyzed. It should, however, be noted that these clusters extend beyond the limits required for promoter function as defined above (Fig. 3), and therefore many more protein bindings sites may be located outside of the promoter sequences that have so far been preferentially investigated. In addition, it should be kept in mind that the protein interaction sites detected by the techniques presently used may only display the stable and less flexible part of DNA/protein and protein/protein interactions. Considering that the distance between any two sequence elements in the viral DNA genome is short relative to the distances of 3–10 kb often observed for single cellular promoter/enhancer regions, the whole 3-kb HBV genome should be considered in toto as a single, complex "regulon" in which regulatory proteins binding to any given *cis*-acting element may productively interact, directly or indirectly, with several other such sites and exert combinatorial effects not detectable with subgenomic DNA elements.

5.2 Regulation of Transcription by Viral Gene Products

Recently it has been shown that the X-gene product is capable of stimulating the activity of a variety of viral and cellular promoters including that of the HBV enhancer and the core gene promoter or the S gene promoter (TWU and SCHLOEMER 1987; SPANDAU and LEE 1988; ZAHM and HOFSCHEIDER 1988; COLGROVE et al. 1989; TWU and ROBINSON 1988; AUFIERO and SCHNEIDER 1990; see SCHEK et al. 1990 for a recent review). The mechanism of this interaction is presently unclear. But interaction of the X-protein with its many targets seems unlikely in view of the very low X-protein concentration in the HBV-infected liver (PFAFF et al. 1987). Instead, direct or indirect modulation of factors regulating viral and cellular gene activity appears to be a more realistic alternative to explain the pleiotropic transactivation properties of the X-gene product. In keeping with this notion, very recent data suggest that the X-protein may possess protein kinase activity (WU et al. 1990), or belong to a family of protease inhibitors (T. TANAKA and K. KOIKE, personal communication). A 3- to 10-fold stimulation of the synthesis of HBsAg and HBcAg and of genomic transcripts by X-transactivation has been observed in differentiated human hepatoma cells during transient expression of HBV genomes lacking a functional X-gene (unpublished results: COLGROVE et al. 1989), and similarly increased levels of activity have been reported for the corresponding promoters if coupled individually to indicator genes such as *CAT* or *LUC* (COLGROVE et al. 1989, RANEY et al. 1990). In the latter experimental setting, strong transactivation was also observed on the enhancer/X promoter element (FAKTOR and SHAUL 1990). Extrapolated to the HBV genome, this suggests that the X-gene product could stimulate its own synthesis. If so, this would obviously require a mechanism for downregulation of X expression after a desired level of X protein has been reached. In agreement with such an autoregulatory scheme, a reduction of HBV gene expression has been observed at high levels of X-gene dosage in transfection experiments (FAKTOR and SHAUL 1990; unpublished data from this laboratory).

Several recent reports suggest that viral gene products encoded outside of the X-gene may also participate in the transcriptional control of viral gene

Saito I, Oya Y, Yamamoto K, Yuasa T, Shimojo H (1985) Construction of nondefective adenovirus type 5 bearing a 2.8-kilobase hepatitis B virus DNA near the right end of its genome. J Virol 54: 711–719

Schek N, Fischer M, Schaller H (1990) The hepadnaviral X protein. In: McLachlan A (ed) Molecular biology of hepatitis B viruses. CRC, Boca Raton (CRC Uniscience, in press)

Schlicht HJ, Salfeld J, Schaller H (1987) The preC region of the duck hepatitis B virus is essential for synthesis and secretion of processed core proteins but not for virus formation. J Virol 61: 3701–3709

Shaul Y (1991) Regulation of hepadnavirus transcription. In: McLachlan A (ed) Molecular biology of hepatitis B viruses. CRC, Boca Raton (CRC Uniscience, in press)

Shaul Y, Ben-Levy R (1987) Multiple nuclear proteins in liver cells are bound to hepatitis B virus enhancer element and its upstream sequences. EMBO J 6: 1913–1920

Shaul Y, Rutter WJ, Laub O (1985) A human hepatitis B viral enhancer element. EMBO J 4: 427–430

Shaul Y, Ben-Levy R, De-Medina T (1986) High affinity binding site for nuclear factor I next to the hepatitis B virus S gene promoter. EMBO J 5: 1967–1971

Simonsen CC, Levinson AD (1983) Analysis of processing and polyadenylation signals of the hepatitis B virus surface antigen gene by using simian virus 40–hepatitis B virus chimeric plasmids. Mol Cell Biol 3: 2250–2258

Spandau DF, Lee CH (1988) Trans-activation of viral enhancers by the hepatitis B virus X protein. J Virol 62: 427–434

Standring DN, Rutter WJ, Varmus HE, Ganem D (1984) Transcription of the hepatitis B surface antigen gene in cultured murine cells initiates within the presurface region. J Virol 50: 563–571

Su TS, Lai CJ, Huang JL, Lin LH, Yauk YK, Chang CM, Lo SJ, Han SH (1989) Hepatitis B virus transcript produced by RNA splicing. J Virol 63: 4011–4018

Suzuki T, Masui N, Kajino K, Saito I, Miyamura T (1989) Detection and mapping of spliced RNA from a human hepatoma cell line transfected with the hepatitis B virus genome. Proc Natl Acad Sci USA 86: 8422–8426

Tognoni A, Cattaneo R, Serfling E, Schaffner W (1985) A novel expression selection approach allows precise mapping of the hepatitis B virus enhancer. Nucleic Acids Res 13: 7457–7472

Treinin M, Laub O (1987) Identification of a promoter element located upstream from the hepatitis B virus X gene. Mol Cell Biol 7: 545–548

Tur-Kaspa R, Burk RD, Shaul Y, Shafritz DA (1986) Hepatitis B virus DNA contains a glucocorticoid-responsive element. Proc Natl Acad Sci USA 83: 1627–1631

Tur-Kaspa R, Shaul Y, Moore DD, Burk RD, Okret S, Poellinger L, Shafritz DA (1988) The glucocorticoid receptor recognizes a specific nucleotide sequence in hepatitis B virus DNA causing increased activity of the HBV enhancer. Virology 167: 630–633

Twu JS, Robinson WS (1989) Hepatitis B virus X gene can transactivate heterologous viral sequences. Proc Natl Acad Sci USA 86: 2046–2050

Twu JS, Schloemer RH (1987) Transcriptional trans-activating function of hepatitis B virus. J Virol 61: 3448–3453

Twu JS, Schloemer RH (1989) Transcription of the human beta interferon gene is inhibited by hepatitis B virus. J Virol 63: 3065–3071

Vannice JL, Levinson AD (1988) Properties of the human hepatitis B virus enhancer: position effects and cell-type nonspecificity. J Virol 62: 1305–1313

Wickens M (1990) How the messenger got its tail: addition of poly(A) in the nucleus. Trends Biochem Sci 15: 277–281

Will H, Reiser W, Weimer T, Pfaff E, Büscher M, Sprengel R, Cattaneo R, Schaller H (1987) Replication strategy of human hepatitis B virus. J Virol 61: 904–911

Wu JY, Zhou ZY, Judd A, Cartwright CA, Robinson WS (1990) The hepatitis B virus-encoded transcriptional trans-activator hbx appears to be a novel protein serine/threonine kinase. Cell 63: 687–695

Yaginuma K, Koike K (1989) Identification of a promoter region for 3.6-kilobase mRNA of hepatitis B virus and specific cellular binding protein. J Virol 63: 2914–2920

Yaginuma K, Shirakata Y, Kobayashi M, Koike K (1987) Hepatitis B virus (HBV) particles are produced in a cell culture system by transient expression of transfected HBV DNA. Proc Natl Acad Sci USA 84: 2678–2682

Yee JK (1989) A liver-specific enhancer in the core promoter region of human hepatitis B virus. Science 246: 658–661

Zahm P, Hofschneider PH, Koshy R (1988) The HBV X-ORF encodes a transactivator: a potential factor in viral hepatocarcinogenesis. Oncogene 3: 169–177

The second model requires the TP for the removal of the r region at the 3′ end of minus-strand DNA and the ligation of the 5′ and 3′ ends following hydrolysis of the protein-DNA bond, a reaction carried out by the A protein of bacteriophage ϕX174 for the circularization of the single-stranded DNA genome (ROTH et al. 1984; BROWN et al. 1984). While the first model primarily depends on cellular DNA modifying enzymes and requires DNA synthesis, the second model relies on an enzymatic activity of the TP and does not require DNA synthesis. The observation that foscarnet, an inhibitor of the viral DNA polymerase, does not block the conversion of virion DNA into CCC DNA suggests that the viral DNA polymerase is not required for the completion of minus-strand DNA (MASON et al. 1987a).

2.2 Amplification and Regulation of CCC DNA Synthesis

TUTTLEMAN et al. (1986b) demonstrated that in primary duck hepatocytes infected in vitro with DHBV, CCC DNA was derived not only from infecting viral DNA as described above but also from molecules synthesized de novo early during infection. De novo synthesized CCC DNA was found to be present 3–4 days after infection at levels estimated to be in 50-fold excess of that derived from the infecting virus. Since no production of infectious virus could be demonstrated in the culture supernatants, it was concluded that the amplification of CCC DNA they observed had occurred through an intracellular pathway not involving infection of new cells, Subsequent studies with an inhibitor of viral attachment and uptake, suramin (PETCU et al. 1988), or with a blocking antiserum provided supporting evidence that a second round of infection was not required for de novo synthesis of CCC DNA (WU et al. 1990). Genetic studies revealed that viral envelope proteins were likewise not required for CCC DNA amplification, thereby indicating that virus assembly was not involved. Several lines of evidence, therefore, support the idea that CCC DNA amplification is an intracellular process.

Amplification of CCC DNA appears to be important for subsequent virus production under some conditions. For example, development of high levels of viral replicative intermediates in infected primary hepatocyte cultures can be inhibited by incorporation of 5-bromodeoxyuridine during CCC DNA amplification (TUTTLEMAN et al. 1986b). In addition, enhanced levels of CCC DNA are associated with a more rapid accumulation of replicative viral DNA in primary

◀ of minus-strand DNA becomes covalently, linked to the terminal protein depicted by the *solid oval symbol*. Pregenomic RNA is degraded during minus-strand DNA synthesis by an RNAseH-like activity (**E**). Plus-strand DNA synthesis initiates at the first nucleotide downstream of DR2 (**F**). It is primed by an ca. 18-nucleotide long RNA primer derived from the 5′ end of pregenomic RNA (**E**). The 3′ end of the primer is determined by a RNAseH activity (indicated by an *open triangle*) prior to its transfer to DR2 (**E**,**F**). Plus-strand DNA synthesis proceeds to the 5′ end of minus-strand DNA, where a template switch to the 3′ end of minus-strand DNA is necessary for the continuation of DNA synthesis (**G**). The template switch is most likely facilitated by the r region at the 5′ and 3′ ends of minus-strand DNA. As a consequence of this reaction, virion DNA assumes a relaxed circular confirmation (**A**) hepatocytes infected with a DHBV genome which fails to suppress amplification (SUMMERS et al. 1990). Whether the effect of amplification is limited simply to supplying more copies of the transcriptional template or whether amplified CCC DNA is transcriptionally different from that derived from infecting viral DNA is not known.

TUTTLEMAN et al. (1986) investigated the mechanism of CCC DNA synthesis during amplification in primary duck hepatocytes congenitally infected with DHBV and maintained in serum-containing medium. Under these culture conditions, levels of CCC DNA were unstable in the infected cells and could increase as much as 100-fold, depending on the cell density, without apparent damage to the cell layer. As in early amplification, this increase was not inhibited by suramin (WU et al. 1990), pointing to the likelihood of an intracellular process similar to that observed at the initiation of infection. Using density-labelling experiments, it was shown that CCC DNA amplification occurred through the reverse transcription pathway, presumably by utilization of DNA-containing nucleocapsids derived from the cytoplasm to augment the nuclear levels of CCC DNA. The same process was thought to occur during early amplification since de novo synthesis of CCC DNA was shown to occur in the absence of any detectable semiconservative replication of CCC DNA derived from the infecting viral DNA (TUTTLEMAN et al. 1986).

Infected cells in culture therefore apparently have the ability to synthesize moderate and sometimes large amounts of CCC DNA in the absence of participation by extracellular virus; yet, in general, the levels of CCC DNA that have been reported to occur in infected liver are relatively low. In a limited survey of ducklings infected with DHBV, under conditions in which every hepatocyte would have been expected to be infected, CCC DNA copy numbers of ca. 3–8 were estimated (PETCU and MASON, unpublished observations). A separate study of a WHV-infected woodchuck revealed a copy number of ca. 50 (J. SUMMERS, unpublished observations). While CCC DNA levels in primary hepatocyte cultures may reach levels of greater than 10^3 per cell in serum-containing medium, culture conditions that more closely maintain the cells in a differentiated state (serum-free medium supplemented with dimethyl sulfoxide) result in the maintenance of low steady state levels of less than 10 copies per infected cell (PUGH and SUMMERS 1989; SUMMERS et al. 1990; J. SUMMERS, unpublished observations). These observations suggest that controls on CCC DNA amplification may limit its levels according to the physiological state of the hepatocyte both in vivo and in vitro.

In a recent study, it was shown that viral envelope protein expression can dramatically suppress CCC DNA amplification in cultured duck hepatocytes maintained in dimethyl sulfoxide-containing medium (SUMMERS et al. 1990). In primary cells infected with a DHBV mutant genome defective in the production of both PreS (P35) and S (P17) envelope proteins, CCC DNA amplification proceeded at an accelerated rate for at least 10 days post infection. In contrast, wild-type genomes showed evidence of suppression of amplification within 4–7 days post infection, and suppression was complete after 7 days. This lack of control exhibited by the mutant genome resulted in the accumulation of 20–30-fold higher levels of CCC DNA than in wild-type infection, normalized to the amount of replicative forms in the infected cultures. Moreover, infection by the mutant virus resulted in only transient expression of DHBV replication in the culture, possibly due to death of the mutant-infected cells. Mutation of the P35 coding region alone has been shown to be sufficient to destroy control of CCC DNA amplification (SUMMERS et al., unpublished observations). The mechanism of suppression of CCC DNA synthesis by P35 is unknown; however, it was found (SUMMERS et al., unpublished observations) that a defect in P35 production also prevents packaging of nucleocapsids into enveloped particles or their secretion from transfected hepatoma cells. This observation suggests that a physical binding of P35 to nucleocapsids may play a role in their assembly into envelopes, thereby preventing their utilization for CCC DNA synthesis in hepatocytes.

It is interesting that the ability of hepadnaviruses to synthesize viral DNA genomes in persistently infected cells enables the infected cells to amplify CCC DNA pools and thereby replenish levels of transcriptional template lost by cell turnover and dilution during cell division. However, amplification is a potentially cytopathic mode of replication in the absence of controls. Apparently, suppression of amplification by envelope proteins is a feedback control for suppressing this cytopathic potential while at the same time maintaining persistent levels of virus production by the cell.

While envelope proteins appear to act at the level of limiting the rate of CCC DNA synthesis, it is not known whether this or other controls can reduce the levels of CC DNA in hepatocytes once high levels are achieved. Reduction of the CCC DNA pool in a nongrowing cell would require degradation or export of the CCC DNA from the nucleus, and this process could potentially lead to partial or complete loss of the ability of an infected cell to transcribe pregenomes and mRNAs and eventually to the loss of virus production. For practical reasons, it will be of great interest to determine whether such down-regulatory controls exist.

3 Pregenomic RNA

3.1 Structure of Pregenomic RNA

CCC DNA is the presumptive template for the transcription of the template for reverse transcription, termed pregenomic RNA (chapter by H. Schaller and M. Fischer, this volume). Although direct evidence is still lacking to confirm this notion, the possibility is further supported by the fact that CCC DNA preparations from cloned viral DNAs are infectious in vivo (WILL et al. 1982; SEEGER et al. 1984; SPRENGEL et al. 1984). Information about the structure of pregenomic RNA is derived from the analysis of viral RNAs isolated from the livers of infected animals, from the characterization of virion DNA, and to some extent from the analysis of RNA isolated from cytoplasmic core particles. Viral RNA isolated from the livers of infected animals contains a ca. 3.5-kb long RNA species and one or two shorter transcripts, ca. 2.1–2.5-kb in length (ENDERS et al. 1985; MOROY et al. 1985; BÜSCHER et al. 1985; chapter by H. SCHALLER and M. FISCHER, this volume). Since replication by reverse transcription requires an RNA template that contains all the information encoded by the viral DNA genome, the 3.5-kb RNA species is the only detectable transcript with the physical structure to function as a replicative intermediate. Since the 3.5-kb transcript has heterogeneous 5′ ends, the nature of pregenomic RNA cannot be inferred solely from the analysis of cytoplasmic RNA. The identity of the 5′ end of pregenomic RNA has been inferred from an analysis of the 5′ ends of plus-strand DNA. These DNA strands are covalently linked to a capped RNA oligomer apparently derived from the 5′ end of one of the major 3.5-kb RNAs (LIEN et al. 1986; SEEGER et al. 1986). As will be discussed below (Sect. 5), since viral DNA synthesis occurs within nucleocapsids, this RNA species appears to be the most frequently packaged template for reverse transcription within the nucleocapsid. Finally, the analogous transcript is found in preparations of cytoplasmic core particles isolated from GSHV-infected livers (ENDERS et al. 1987; LAVINE et al. 1989).

The synthesis of pregenomic RNA leads to a product with terminally redundant ends (R) that exceeds the size of its CCC DNA template by the length of exactly one copy of the terminal repetition (BÜSCHER et al. 1985; ENDERS et al. 1985; chapter by H. SCHALLER and M. FISCHER, this volume; Fig. 1D). An 11–12-nucleotide long sequence termed direct repeat 1 (DR1) is located at both the 5′ end and near the 3′ end of pregenomic RNA. A third copy of the repeat, termed DR2, is located upstream of DR1 close to the 3′ end of pregenomic RNA. As will be discussed below, DR1 and DR2 bear signals important for the initiation of minus- and plus-strand DNA synthesis.

The biological activity of this RNA species to direct viral DNA synthesis has indirectly been validated by experiments in which a cDNA equivalent of pregenomic RNA of HBV and WHV has been fused to heterologous promoters (JUNKER et al. 1987; SEEGER et al. 1989). Transfection of such constructs into cultured cells can lead to the production of infectious virus particles. Most importantly, with these constructs expression of hepadnaviruses is no longer restricted to hepatoma cells. The expression of WHV in fibroblasts leads to the production of infectious virus as determined by the inoculation of susceptible woodchucks (SEEGER et al. 1989). From these experiments it can be inferred that the steps in the hepadnaviral replication cycle that follow the synthesis of pregenomic RNA do not depend on hepatocyte-specific functions.

3.2 Packaging of Pregenomic RNA

To synthesize the viral genome, pregenomic RNA is packaged along with the viral polymerase into core particles. This process most likely requires the interaction of *cis*-acting sequences on pregenomic RNA with components on the core subunits. Alternatively, it is possible that pregenomic RNA and polymerase are encapsidated as a complex where the polymerase bears the signals for packaging (Sect. 6).

ENDERS et al. provided evidence that the 5′ end of GSHV RNA provides signals for encapsidation by demonstrating that the shortest of the three 3.5-kb transcripts, pregenomic RNA, is predominantly packaged into core particles (ENDERS et al. 1987). However, it now appears that the packaging of the longer transcripts is suppressed as a consequence of the translation of the precore gene. HBV and WHV variants with mutations in the initiation codon of the precore gene can package the longer 3.5-kb RNA transcripts (JUNKER et al. 1987; C. SEEGER, unpublished results), suggesting that translational usage of the RNA plays a role in whether it is packaged into nucleocapsids. Recently, M. NIEPMANN-JUNKER et al. reported that a hybrid RNA transcript composed of sequences derived from HBV and the β-galactosidase gene is incorporated into cytoplasmic core particles (NIEPMANN-JUNKER et al. 1990). The HBV fragment in this construct corresponded to the 5′ end of pregenomic RNA including the precore and a portion of the core regions. This result suggests that sequences near the 5′ end of this hybrid RNA, but found at both the 5′ and 3′ ends of the pregenomic RNA, bear the signals for the packaging of the viral RNA. A detailed genetic analysis of these regions on the pregenome should reveal the sequence motif(s) important for its encapsidation.

4 Minus-Strand DNA Synthesis

4.1 Template for Minus-Strand DNA Synthesis

Infected livers contain partially duplexed and free minus-strand DNA species of varying length, ranging from ca. 3-kb to a few hundred bases (MASON et al. 1982; WEISER et al. 1983). A similar pattern is observed when DNA is isolated from viral core particles purified from the cytoplasm of infected hepatocytes (SUMMERS and MASON 1982). Moreover, the incubation of core particles with Mg^{2+} and nucleotide triphosphates leads to the extension of minus- and plus-strand DNA, indicating that core particles contain a DNA polymerase activity similar to the one identified in virions purified from sera and that core particles are the site of viral DNA synthesis (SUMMERS and MASON 1982). At least three observations indicate that RNA is the template for the synthesis of minus-strand DNA. First, the incubation of core particles with actinomycin D, an inhibitor of DNA-directed DNA synthesis, inhibits plus-strand DNA synthesis but does not interfere with the extension of minus strands. Second, the sedimentation analysis of the products obtained from reactions in the presence of actinomycin D reveals that minus-strand DNA is base-paired with RNA. Third, during minus-strand DNA synthesis

Molnar-Kimber KL, Summers JW, Taylor JM, Mason WS (1983) Protein covalently bound to minus strand DNA intermediates of duck hepatitis B virus. J Virol 45: 165–172

Molnar-Kimber KL, Summers JW, Mason WS (1984) Mapping of the cohesive overlap of duck hepatitis B virus DNA and of the site of initiation of reverse transcription. J Virol 51: 181–191

Moroy T, Etiemble J, Trepo C, Tiollais P, Buendia MA (1985) Transcription of woodchuck hepatitis virus in the chronically infected liver. EMBO J 4: 1507–1514

Niepmann-Junker M, Bartenschlager R, Schaller H (1990) A short cis-acting sequence is required for hepatitis B virus pregenome encapsidation and sufficient for packaging of foreign RNA. EMBO 9: 3389–3396

Ochiya T, Tsurimoto T, Ueda K, Okubo K, Shiozawa M, Matsubara K (1989) An *in vitro* system for infection with hepatitis B virus that uses primary human fetal hepatocytes. Proc Natl Acad Sci USA 86: 1875–1879

Pasek M, Goto T, Gilbert W, Zink B, Schaller H, MacKay P, Leadbetter G, Murray K (1979) Hepatitis B virus genes and their expression in *E coli*. Nature 282: 575–579

Petcu DJ, Aldrich CE, Coates L, Taylor JM, Mason WS (1988) Suramin inhibits *in vitro* infection by duck hepatitis B virus, Rous sarcoma virus, and hepatitis delta virus. J Virol 167: 385–392

Pugh JC, Yaginuma K, Koike K, Summers J (1988) Duck hepatitis B virus (DHBV) particles produced by transient expression of DHBV DNA in a human hepatoma cell line are infectious *in vitro*. J Virol 62: 3513–3616

Pugh JC, Summers J (1989) Infection and uptake of duck hepatitis B virus by duck hepatocytes maintained in the presence of dimethyl sulfate. J Virol 172: 564–572

Radziwill G, Zentgraf H, Schaller H, Bosch V (1988) The duck hepatitis B virus polymerase is tightly associated with the viral core structure and unable to switch to an exogenous template. Virology 163: 123–132

Radziwill G, Tucker W, Schaller H (1990) Mutational analysis of the hepatitis B virus P gene product domain structure and RNAseH activity. J Virol 64: 613–620

Rijnitjes PJM, Morshage HJ, Yap SH (1988) *In vitro* infection of primary cultures of cryopreserved adult human hepatocytes with hepatitis B virus. Virus Res 10: 95–110

Robinson WS, Clayton DA, Greenman RL (1974) DNA of a human hepatitis B virus candidate. J Virol 14: 384–391

Roth MJ, Brown DR, Hurwitz J (1984) Analysis of bacteriophage phiX174 gene A protein-mediated termination and reinitiaion of phiX DNA synthesis. J Biol Chem 259: 10556–10568

Rothberg PG, Harris TJ, Nomoto A, Wimmer E (1978) The genome-linked protein of picornaviruses, V. O^4-(5'-uridylyl) tyrosine is the bond between the genome-linked protein and the RNA of poliovirus. Proc Natl Acad Sci USA 75: 4868–4872

Roychoudhury S, Shih C (1990) cis Rescue of a mutated reverse transcriptase of human hepatitis B virus by creation of an internal AUG. J Virol 64: 1063–1069

Ruiz-Opazo N, Chakraborty PR, Shafritz DA (1982) Evidence for supercoiled hepatitis B virus DNA in chimpanzee liver and serum Dane particles: possible implications in persistent HBV infection. Cell 29: 129–138

Salas M (1983) A new mechanism for the initiation of replication of ϕ29 and adenovirus DNA: priming by the terminal protein. In: Doerfler W (ed) The molecular biology of adenoviruses 1. 30 years of adenovirus research 1953–1983. Springer, Berlin Heidelberg New York, pp 89–106 (Current topics in microbiology and immunology, vol 109)

Sattler F, Robinson WS (1979) Hepatitis B viral DNA molecules have cohesive ends. J Virol 32: 226–233

Schlicht HJ, Radziwill G, Schaller H (1989) Synthesis and encapsidation of duck hepatitis B virus reverse transcriptase do not require formation of core-polymerase fusion proteins. Cell 56: 85–92

Seeger C, Maragos J (1989) Molecular analysis of the function of direct repeats and a polypurine tract for plus-strand DNA priming in woodchuck hepatitis virus. J Virol 63: 1907–1915

Seeger C, Maragos J (1990) Identification and characterization of the woodchuck hepatitis virus origin of DNA replication. J Virol 64: 16–23

Seeger C, Ganem D, Varmus HE (1984) The cloned genome of ground squirrel hepatitis virus is infectious in the animal. Proc Natl Acad Sci USA 81: 5849–5852

Seeger C, Ganem D, Varmus HE (1986) Biochemical and genetic evidence for the hepatitis B virus replication strategy. Science 232: 477–484

Seeger C, Baldwin B, Tennant BC (1989) Expression of infectious woodchukc hepatitis virus in murine and avian fibroblasts. J Virol 63: 4665–4669

Sells MA, Chen M, Acs G (1987) Production of hepatitis B virus particles in hepG2 cells transfected with cloned hepatitis B virus DNA. Proc Natl Acad Sci USA 84: 1005–1009

Sells MA, Zelent AZ, Shvartsman M, Acs G (1988) Replicative intermediates of hepatitis B virus in HepG2 cells that produce infectious virions. J Virol 62: 2836–2844

Shabarova ZA (1970) Synthetic nucleotide-peptides. In: Davidson JN, Cohn W (eds) Progress in nucleic acid research and molecular biology, vol. X. Academic, New York, pp 145–182

Shih C, Li L-S, Roychoudhury S, Ho M-H (1989) *In vitro* propagation of human hepatitis B virus in a rat hepatoma cell line. Proc Natl Acad Sci USA 86: 6323–6327

Sprengel R, Kuhn C, Manso K, Will H (1984) Cloned duck hepatitis B virus DNA is infectious in pekin ducks. J Virol 52: 932–937

Sprengel R, Kaleta EF, Will H (1988) Isolation and characterization of a hepatitis B virus endemic in herons. J Virol 62: 3832–3839

Stillmann B (1989) Initiation of eukaryotic DNA replication *in vitro*. Annu Rev Cell Biol 5: 197–246

Summers J, Mason WS (1982) Replication of the genome of a hepatitis B-like virus by reverse transcription of an RNA intermediate. Cell 29: 403–415

Summers J, O'Connell A, Millman I (1975) Genome of hepatitis B virus restriction enzyme cleavage and structure of DNA extracted from Dane particles. Proc Natl Acad Sci USA 72: 4597–4601

Summers J, Smolec JM, Snyder R (1978) A virus similar to human hepatitis B virus associated with hepatitis and hepatoma in woodchucks. Proc Natl Acad Sci USA 75: 4533–4537

Summers J, Smith PM, Horwich AL (1990) Hepadnaviral envelope proteins regulate covalently closed circular DNA amplification. J Virol 64: 2819–2824

Sureau C, Romet-Lemonne J, Mullins JI, Essex M (1987) Production of hepatitis B virus by a differentiated human hepatoma cell line after transfection with cloned circular HBV DNA. Cell 47: 37–47

Sureau C, Eichberg JW, Hubbard GB, Romet-Lemonne JL, Essex M (1988) A molecularly cloned hepatitis B virus produced *in vitro* is infectious in a chimpanzee. J Virol 62: 3064–3067

Tagawa M, Omata M, Okuda K (1986) Appearance of viral RNA transcripts in the early stage of duck hepatitis B virus infection. Virology 152: 477–482

Takeda N, Kuhn RJ, Yang C, Takegami T, Wimmer E (1986) Initiation of poliovirus plus-strand RNA synthesis in a membrane complex of infected HeLa cells. J Virol 60: 43–53

Tanese N, Roth M, Goff SP (1985) Expression of enzymatically active reverse transcriptase in *Echerichia coli*. Proc Natl Acad Sci USA 82: 4944–4948

Tanese N, Sodrowski J, Haseltine WA, Goff SP (1986) Expression of reverse transcriptase activity of human T-lymphotropic virus type III (HTLV-III/LAV) in *Escherichia coli*. J Virol 59: 743–745

Taylor JM, Illmensee R (1975) Site on the RNA of an avian sarcoma virus at which primer is bound. J Virol 16: 533–558

Theze N, Gripon P, Fourel I, Hantz O, Trepo C, Guguen-Guillouzo C (1987) Maintenance of woodchuck hepatitis virus activity in primary cultures of woodchuck hepatocytes. J Gen Virol 68: 1029–1039

Tobin GJ, Young DC, Flanegan JB (1989) Self-catalyzed linkage of poliovirus terminal protein VPg to poliovirus RNA. Cell 59: 511–519

Toh H, Hayashida H, Miyata T (1983) Sequence homology between retroviral reverse transcriptase and putative polymerases of hepatitis B virus and cauliflower mosaic virus. Nature 305: 827–829

Tsurimoto T, Fujiyama A, Matsubara K (1987) Stable expression and replication of hepatitis B virus genome in an integrated state in a human hepatoma cell line transfected with the cloned viral DNA. Proc Natl Acad Sci USA 84: 444–448

Tuttleman JS, Pugh JC, Summers JW (1986a) *In vitro* experimental infection of primary duck hepatocyte cultures with duck hepatitis B virus. J Virol 58: 17–25

Tuttleman JS, Pourcel C, Summers JW (1986b) Formation of the pool of covalently closed circular viral DNA in hepadnavirus-infected cells. Cell 47: 451–46

Varmus HE, Swanstrom R (1984) Replication of retroviruses. In: Weiss R, Teich N, Varmus HE, Coffin J (eds) RNA tumor viruses. Cold Spring Harbor Laboratory, Cold Spring Harbor, pp 369–512

Varmus HE, Swanstrom R (1985) Replication of retroviruses. In: Weiss R, Teich N, Varmus HE, Coffin J (eds) RNA tumor viruses. Cold Spring Harbor Laboratory, Cold Spring Harbor, pp 75–134

Weiser B, Ganem D, Seeger C, Varmus HE (1983) Closed circular viral DNA and asymmetric heterogeneous forms in livers from animals infected with ground squirrel hepatitis virus. J Virol 48: 1–9

Will H, Cattaneo R, Koch H, Darai G, Schaller H, Schellekens H, van Eer P, Deinhardt F (1982) Cloned HBV DNA causes hepatitis in chimpanzees. Nature 299: 740–742

Will H, Reiser W, Weimer T, Pfaff E, Büscher M, Sprengel R, Cattaneo R, Schaller H (1987) Replication strategy of human hepatitis B virus. J Virol 61: 904–911

Wimmer E (1982) Genome-linked proteins of viruses. Cell 28: 199–201

Wu T-T, Coates L, Aldrich CE, Summers J, Mason WS (1990) In hepatocytes infected with duck hepatitis B virus, the template for viral RNA synthesis is amplified by an intracellular pathway. Virology 175: 255–261

Yaginuma K, Shirakata Y, Kobayashi M, Koike K (1987) Hepatitis B virus (HBV) particles are produced in a cell culture system by transient expression of transfected HBV DNA. Proc Natl Acad Sci USA 84: 2678–2682

Assembly of Hepadnaviral Virions and Subviral Particles

D. GANEM

One of the most distinctive features of hepadnaviral infection is the production by infected cells of multiple types of virus-related particles, including large numbers of noninfections subviral particles as well as complete virions. Although this unusual property was among the first aspects of the viral life cycle to be discovered, it remains one of its least well understood aspects. Why do infected cells produce subviral particles? What factors determine whether virions or subviral particles are produced, and how are their relative levels controlled? What are the cellular pathways involved in the production of each particle type? At present our answers to all of these questions are fragmentary. However, the recent application of biochemical and genetic approaches to the study of hepadnaviral morphogenesis has reawakened interest in this process and helped to frame the overall dimensions of the problem. In the following pages I review what is known about hepadnaviral structure and assembly, with emphasis on recent mutational approaches that examine the role of individual viral gene products in the assembly process. Since the work is most advanced for the human hepatitis B virus (HBV), this review will concentrate on HBV; mention will be made of the animal hepadnaviruses only where significant differences exist or where the corresponding information for HBV is not available.

Departments of Microbiology and Immunology and Medicine, University of California Medical Center, San Francisco, CA 94143-0503, USA

Current Topics in Microbiology and Immunology, Vol. 168

1 Hepadnavirus Structure and Replication: An Overview

Shortly after the discovery of the so-called Australia antigen in the serum of multiply transfused individuals (BLUMBERG et al. 1965), it was found that the immunoreactive material resided in a 20-nm virus-like particle (BAYER et al. 1968; GERIN et al. 1969). For a brief period following its discovery, this particle was mistakenly assumed to be the infectious virion of HBV itself. Soon thereafter, however, electron-microscopic examinations of partially purified, serum-derived preparations of HBV revealed the presence of three morphologically distinct particle types: spheres and filaments of 20-nm diameter and less abundant 42-nm double-shelled particles now known as "Dane particles" (DANE et al. 1970). Immunologic studies demonstrated that all three particle types share on their surface the Australia antigen (now called hepatitis B surface antigen, HBsAg) and hence are clearly of viral origin.

The Dane particle is now known to be the infectious virion of HBV (ROBINSON and LUTTWICK 1976). It consists of an outer shell or envelope comprised of HBsAg, presumably in association with host-derived lipid. Internal to this is the viral nucleocapsid or core, a $T=3$ icosahedral structure composed of 180 subunits of a single polypeptide known as hepatitis B core antigen (HBcAg) (S. Fuller and M. Nassal, personal communication). Within the core is the viral genome, partially duplex 3.2-kb relaxed circular DNA molecule, and a polymerase centrally involved in its replication (ROBINSON and GREENMAN 1974; SUMMERS et al. 1975). Titres of Dane particles in the circulation are extremely variable from individual to individual (and often within a single individual over time), ranging from 10^3 to 10^9 per milliliter of blood.

The 20-nm particles initially visualized in HBV preparations are now recognized to be subviral particles that are composed solely of HBsAg and host-derived lipid (PETERSON 1981; SHIH and GERIN 1977; AGGERBECK and PETERSON 1985). Their lipids consist mainly of phospholipids (notably phosphatidyl choline), free and esterfied cholesterol, and small quantities of triglycerides (GAVILANES et al. 1982). They contain no viral genetic information and are therefore noninfectious; they also lack nucleocapsid components and have no detectable cellular polypeptides. They are enormously abundant in the bloodstream of infected carriers, typically exceeding Dane particles in number by a factor of 10^3–10^5. Paradoxically, they are also highly immunogenic: in most individuals they elicit a strong antibody response that is capable of conferring immunity to reinfection in vivo. For this reason, 20-nm particles purified from infected serum (or generated by recombinant DNA technology, see below) have been exploited as a subunit vaccine for HBV infection (SZMUNESS et al. 1980; VALENZUELA et al.1982; MCALEER et al. 1984; WAMPLER et al. 1985).

The role of these particles in the natural history of infection is unknown, but an attractive speculation is that they serve to adsorb neutralizing antibodies and thus facilitate the progression of infection by virions. That such adsorption occurs in vivo is clearly attested to by the frequent development of arthritis due to the deposition of HBsAg/anti-HBs immune complexes in the joints of acutely

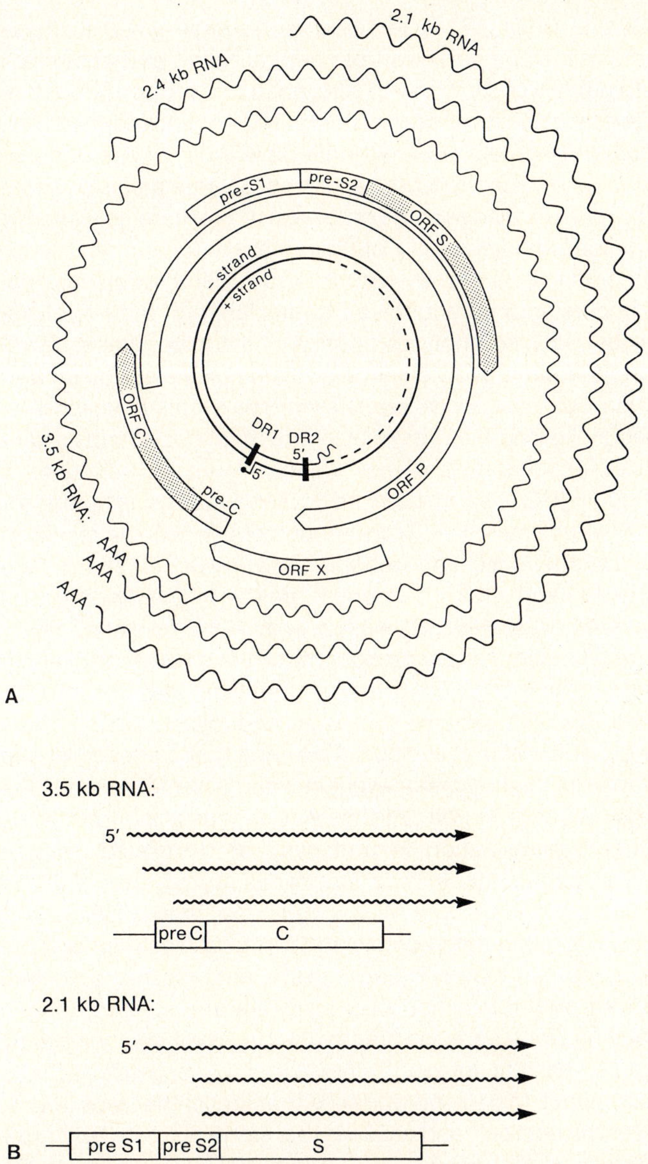

Fig. 1. A The HBV genome. *Wavy lines*, viral transcripts; *boxes*, viral open reading frames (ORFs); *arrows*, direction of transcription and translation; *innermost circles*, structure of virion DNA; *DR1*, *DR2*, positions of the direct repeat sequences involved in the priming steps in viral DNA synthesis. **B** Fine structure of the 5′ ends of the core (*top*) and 2.1-kb surface (*bottom*) mRNAs, relative to the translation starts of the preC and C coding regions (*top*), and preS2 and S coding regions (*bottom*). (From GANEM and VARMUS 1987)

infected patients (GOCKE 1975). What remains at issue is whether these complexes do indeed shield Dane particles from neutralization and accelerate horizontal spread within the liver. One way to approach this problem would be to examine viral mutants that can synthesize virions but not subviral particles. Unfortunately, attempts to generate such mutants have thus far been unsuccessful (V. BRUSS and D. GANEM, unpublished results), a fact that points out both the rudimentary state of our knowledge of the pathways of viral assembly and the potential utility of a deeper understanding of this area.

Since HBV morphogenesis is closely coupled to other aspects of viral replication, such an understanding requires an appreciation of the overall life cycle of hepadnaviruses. Detailed analyses of each of the major steps in viral replication can be found in other chapters in this volume (see also GANEM and VARMUS 1987); here I provide only a brief overview of the events in which viral assembly is embedded. Following the uptake of virions, viral nucleocapsids are transported to the cell nucleus, where their genomes are converted to a covalently closed circular (CCC) form believed to serve as the template for transcription by host RNA polymerase II. Transcription of the genome results in the formation of both subgenomic and genomic mRNAs whose structure is depicted in Fig. 1. The 2.1- and 2.4-kb transcripts are the mRNAs for the viral surface antigens (CATTANEO et al. 1983; STANDRING et al. 1984; WILL et al. 1987). The genomic RNAs are bifunctional, serving not only as the mRNAs for the core proteins (and for the polymerase) but also as the templates for reverse transcription back into the DNA genome (SUMMERS and MASON 1982; ENDERS et al. 1985; BÜSCHER et al. 1985; WILL et al. 1987). Following their transport to and translation in the cytoplasm, the first major event in virion assembly occurs: the selective encapsidation of genomic RNA into the nucleocapsid, along with the viral polymerase. Once encapsidation is underway or completed, reverse transcription of the RNA begins, generating first minus strand, then partially duplex DNA intermediates (SUMMERS and MASON 1982; LIEN et al. 1986; SEEGER et al. 1986). Nucleocapsids containing completed DNA genomes then acquire their outer envelope, presumably by budding into the endoplasmic reticulum (ER) in areas derivatized by the insertion of transmembrane surface antigen polypeptides. The resulting Dane particles are then transported from the cell by the normal pathways of vesicular transport. Viral assembly and release thus occur without cell lysis. In fact, there is no evidence that any of the steps in viral replication is injurious to the host cell, and cell lines that stably produce HBV grow normally in culture (SUREAU et al. 1986; TSURIMOTO et al. 1987; SELLS et al. 1987).

2 Core Particle Formation

Nucleocapsid formation involves the correct assembly of three viral constituents: core antigen, the polymerase and the genomic RNA template. No major cellular proteins have been detected in the core but a role for host factors in nucleocapsid assembly has by no means been excluded. In this section we examine how each of these viral components participates in the assembly process.

2.1 Core Antigens

Core polypeptides are encoded by the C open reading frame (ORF C, Fig. 1; BURRELL et al. 1979; PASEK et al. 1978). Examination of the sequence of this coding region reveals the existence of two in-frame initiation codons (GALIBERT et al. 1979; VALENZUELA et al. 1980). The major nucleocapsid protein originates from the second of these AUGs. To allow ribosomes access to this internal AUG, the viral transcriptional program specifies multiple 5′ ends for the genomic RNAs that serve as the C gene messages (cf. Fig. 1B; ENDERS et al. 1985; WILL et al. 1987; YAGINUMA et al. 1987). The most abundant of these begins between the two initiators, thereby positioning the internal AUG as the first initiator on the transcript, a configuration favorable for efficient initiation of translation (KOZAK 1989). Heterologous viral vectors expressing ORF C from this AUG codon produce a cytoplasmic 21-kDa protein that comigrates with the authentic polypeptide purified from serum-derived virions (WILL et al. 1984; ROOSINCK et al. 1986; MCLACHLAN et al. 1987).

Core antigen has a number of properties in common with the nucleocapsid proteins of other enveloped viruses. First, it is a phosphoprotein: incubation of partially purified cores with gamma-[32P]ATP leads to its phosphorylation on serine residues (ALBIN and ROBINSON 1980; GERLICH et al. 1982; FEITELSON et al. 1982). Similar labeling occurs when transfected cells expressing HBcAg are metabolically labeled with [^{32}P]-orthophosphate (ROOSINCK and SIDDIQUI 1987). In duck HBV (DHBV), and most likely in HBV as well, most of the phosphorylation is at the C-terminus of the protein (SCHLICHT et al. 1989). The predicted amino acid sequence of core antigen does not contain sequences homologous to the kinase domains of known serine/threonine kinases, and HBcAg produced in *Escherichia coli* is unphosphorylated, arguing that the activity is not intrinsic to the C gene product itself (SCHLICHT et al. 1989). Most evidence suggests that the responsible kinase is of host origin, although its precise identity is uncertain. The function of core phosphorylation, if any, is uncertain, but it is noteworthy that this feature has been conserved in all of the animal hepadnaviruses (FEITELSON et al. 1982; SCHLICHT et al. 1989; PUGH et al. 1989). A recent examination of DHBV infection noted that cores from serum-derived virions were underphosphorylated relative to cytoplasmic cores purified from infected cells (PUGH et al. 1989); this raises the interesting possibility that core dephosphorylation might be involved in signaling mature cores for envelopment or export.

Another striking feature of HBcAg is that the C-terminal portion of the molecule is extremely basic, with arginines heavily overrepresented. This suggested early on that the protein would display nonspecific nucleic acid binding activity, an inference that has been sustained by a variety of assays using natural or recombinant HBcAg and labeled DNA or RNA (PETIT and PILLOT 1985; GALLINA et al. 1989). Removal of the C-terminal region abolishes this binding (GALLINA et al. 1989), although a report to the contrary (MATSUDA et al. 1988) has appeared. Heterologous nucleic acids are bound as well as homologous ones, and there is no evidence as yet for sequence specificity in the protein: nucleic acid interactions being studied in these assays. In recent experiments with DHBV, deletions in the C-terminus of the core protein affected neither particle formation nor genomic RNA packaging, but the resulting particles were noninfectious (SCHLICHT et al. 1989).

This is not to say that core polypeptides might not play a role in the selective encapsidation of genomic RNA during nucleocapsid assembly. Such is the case for retroviral *gag* proteins (MERIC and SPAHR 1986), though there is no conservation in hepadnaviruses of the reiterated *cys-his* sequence motifs implicated in this function in retroviruses. Support for such a role, however, must await the development of binding assays in which appropriate target sequences of HBV RNA (see below) are selectively bound in vitro, or the identification of mutants in ORF C that are assembly competent but encapsidation defective. No such mutants have yet been generated.

Core polypeptides are able to assemble into nucleocapsids in the absence of all other viral proteins. This is most clearly demonstrated by the observation that recombinant vectors expressing ORF C in hosts as divergent as COS cells, yeast, and *Escherichia coli* all direct the formation of morphologically correct nucleocapsids (COHEN and RICHMOND 1982; UY et al. 1986; MIYANOHARA et al. 1986; OU et al. 1986). Such assembly is unimpeded by the addition of heterologous sequences to the N-terminus of the chain (CLARKE et al. 1987; STAHL and MURRAY 1989). From the results in *E. coli* it is sometimes inferred that no specialized host proteins are also required, but a role for evolutionarily conserved functions like heat shock proteins or chaperonins—known to be important in other protein oligomerization processes (HEMMINGSEN et al. 1988)—is not excluded.

Another important implication of these results is that pregenomic RNA is not required to nucleate proper core assembly, since this transcript is not present in these systems. Interestingly, nuclease protection experiments indicate that the *E. coli*-produced cores are not entirely devoid of nucleic acid. Rather, they appear to contain moderate amounts of RNA that may be enriched for sequences of viral origin (BIRNBAUM and NASSAL 1990; V. Bruss, personal communication). The significance of this observation is uncertain; the suggested enrichment could simply be due to the fact that HBcAg mRNA may be the most abundant transcript in the cells at the time of assembly. Certainly other viral proteins known to play a key role in encapsidation (see below) are absent in these systems, and conclusions about the role of core polypeptides in genomic RNA selection should be drawn from these experiments with caution.

One last feature of ORF C deserves mention. Recall that this coding region contains two in-frame AUG codons, with the second one used for HBcAg expression. It turns out that the first initiator is also used for translational initiation in vivo, with the longer genomic transcripts that retain this AUG serving as message (Fig. 1B). As OU et al. (1986) were the first to point out, expression from this initiator confers new biologic properties on the polypeptide. The translation of the so-called preC region between the AUGs generates a signal sequence for protein translocation into the ER. From there, the chains travel through the secretory pathway, undergoing cleavage of their C-terminal basic domain in the process. This generates a series of secreted 16–18 kDa proteins with altered antigenicity, now termed hepatitis B e antigen (HBeAg; OU et al. 1986; MCLACHLAN et al. 1987; STANDRING et al. 1988). In infected patients these proteins can be detected in the bloodstream; where they are correlated with elevated titres of circulating Dane particles and correspondingly high infectivity (OKADA et al. 1976; GANEM 1982).

The processing protease is likely a cellular enzyme that is (partially) inhibited by high concentrations of pepstatin, suggesting it may be an aspartyl protease (JEAN-JEAN et al. 1989a). Suggestions that the protease is of viral origin, stemming from the recognition of regions of sequence homology between the N-terminal region of ORF C and known eukaryotic aspartyl proteases (MILLER 1987), have not been supported by genetic studies: mutational ablation of the putative active sites of this protease does not impair the processing (NASSAL et al. 1989; JEAN-JEAN et al. 1989b).

The ER translocation reaction displays several unusual properties, one of which is its inefficiency: many pre-C chains of woodchuck hepatitis virus (WHV) are apparently not translocated at all, judging from the failure of their signal sequence to be removed (WEIMER et al. 1987). This is highly unusual, since for most secreted proteins nascent chains are cotranslationally translocated across the ER bilayer and their signal sequences removed even before synthesis of the polypeptide is completed. Even more remarkable is the observation that, following partial translocation into the ER lumen and cleavage of the signal sequence by signal peptidase, chain translocation is frequently aborted and the N-terminally processed chain "regurgitated" back into the cytoplasm (GARCIA et al. 1988). Strikingly, recent experiments of OU et al. (1989) indicate that many of the processed chains can also be transported to the nucleus! It is unclear whether the nuclear material derives from the translocated or the regurgitated population of processed chains, though the latter seems more likely. Studies of deletion mutants in the preC region implicate the ten preC amino acid residues between the signal peptide cleavage site and the C AUG as essential to nuclear transport. The important nuclear translocation signals reside in the basic C-terminus of the core proteins; perhaps these preC sequences affect the conformation or accessibility of these signals (YEH et al. 1990). These experiments help to explain the long-standing observation of the nuclear localization of HBcAg reactivity in immunofluorescence studies (HOOFNAGLE et al. 1975, 1978). The function of this nuclear transport reaction in the life cycle is unproven, though it is tempting to believe that it has something to do with the nuclear delivery of encapsidated DNA genomes for conversion to CCC

DNA (TUTTLEMAN et al. 1986). However, this will require proof that the regurgitated form of core antigen (a) is assembled into nucleocapsids containing viral DNA; and (b) can serve as the precursor of the nuclear HBcAg. In addition, the fact that mutants deficient in this protein are viable (see below) means that this cannot be the only nuclear delivery pathway for cores.

The observation that preC translation targeted C-related proteins to the ER, together with the known clinical correlation of serum HBeAg with infectivity, initially suggested that precore proteins might play an essential role in virion morphogenesis (OU et al. 1986; BRUSS and GERLICH 1988). The fact that the preC region is conserved in all animal hepadnaviruses (KODAMA et al. 1985) made this speculation still more attractive. However, mutational ablation of preC translation in DHBV was soon shown to have no effect on viral replication, assembly, and infectivity in both cell culture and the intact animal host (CHANG et al. 1987; SCHLICHT et al. 1987b). Subsequent studies have detected naturally occurring HBV preC-deficient mutants in the circulation of some HBV carriers as well (BRUNETTO et al. 1989; OKAMOTO et al. 1990). To this day, the function of e antigen remains in enigma. SCHLICHT and SCHALLER (1989) have recently shown that HBeAg determinants can be expressed on the surface of cells infected with recombinant vaccinia vectors expressing this protein, and that this expression does not result from the nonspecific adsorption of secreted HBeAg to the cell membrane. Perhaps it is delivered to the surface by pathways akin to those involved in the presentation of endogenous antigens to cytotoxic T cells, via an association of peptide fragments with class I major histocompatibility (MHC) molecules (TOWNSEND and BODMER 1989). If so, then e antigen may be important in triggering cytotoxic host immune responses to HBV that may play a role in disease pathogenesis or viral clearance. But these activities would be expected to restrict or curtail virus spread. So the paradox remains: Why has expression of pre-C been retained in viral evolution?

2.2 Encapsidation of Genomic RNA

The assembly of a complete nucleocapsid requires that the pregenomic RNA that is the template for reverse transcription be packaged, along with the viral polymerase. RNA packaging is selective: detailed studies of the viral RNA within cores of ground squirrel hepatitis virus (GSHV) and DHBV indicates that genomic but not subgenomic RNAs are encapsidated (ENDERS et al. 1987; HIRSCH et al. 1990). The mechanism of selection appears to be stringent. Even among the family of genomic RNAs, whose members differ by only 31nt owing to the heterogeneous placement of the 5′ end (cf. Fig. 1B), only a single RNA—the shortest—is packaged. Since the longer genomic RNAs contain all of the sequences present in the shortest, it is clear that sequence alone cannot entirely account for this discrimination. Recent studies from two groups have shed light on this interesting step in assembly, with regard to both the proteins involved in RNA selection and the nature of the *cis*-acting recognition elements.

In both DHBV (HIRSCH et al. 1990) and HBV (BARTENSCHLAGER et al. 1990; R. HIRSCH, unpublished data), viruses with null mutations in the polymerase gene (ORFP) are unable to encapsidate genomic RNA, producing viral cores with no viral RNA. In the DHBV case, the inability to package genomic transcripts did not lead to enhanced encapsidation of other viral RNAs, as might have been expected if core antigen had a limited ability to carry out this step. The results indicate that the P protein has an essential role in the encapsidation reaction, though the mechanistic details of this role are unknown. Among the possibilities are that the protein interacts directly with specific sequences on the RNA, or that it forms a noncovalent complex with other proteins (e.g., core antigen) which then allows binding of the complex to the transcript; alternatively, the polymerase might initiate DNA synthesis on free pregenomic RNA, with the nascent RNA:DNA hybrid forming the packaging signal that is recognized by core proteins. Of these models, only the last can be excluded at present, since missense mutations in the P gene that destroy the ability of the resulting protein to carry out DNA synthesis nonetheless package RNA at normal efficiency (HIRSCH et al. 1990; BARTENSCHLAGER et al. 1990).

A curious feature of the involvement of P gene products in packaging is that it is *cis*-preferential. That is, in most cases in which wild-type genomes are used to complement P^- mutants, the encapsidated progeny genomes are predominantly of wild-type origin (LAVINE et al. 1989; HIRSCH et al. 1990; BARTENSCHLAGER et al. 1990). The mechanism underlying this is not understood. It could simply reflect the limiting quantities of P proteins in the cell, leading to a preference for action on RNAs nearest to their site of synthesis; alternatively, there could be a preference for nascent P chains to interact with their own genomic mRNA while still on polysomes. However, *cis*-packaging is clearly not obligatory. A low level of *trans*-complementation does occurs, and recombinant vectors designed to overexpress P proteins complement P^- mutants substantially more efficiently (T. WU, L. CONDREAY, and W. MASON, unpublished results; BARTENSCHLAGER et al. 1990). This indicate that there is no mechanistic barrier to RNA recognition and packaging by mature released chains.

Similarly rapid progress has been made in defining the *cis*-acting elements that earmark the genome for encapsidation. JUNKER-NIEPMANN et al. (1990) have shown that deletions originating within the preC region of HBV are packaging defective, and that transfer of sequences spanning this region to the 5′ end of a heterologous (lacZ) mRNA allowed encapsidation of the latter in HBV cores. Likewise, in the DHBV system, preC linker insertions and deletions abolish encapsidation (R. HIRSCH and D. GANEM, submitted). The involved regions of both viruses are present in the terminal redundancy of the genomic RNA, raising the question of which of the repeated copies is recognized by the packaging machinery. The study of DHBV mutants in which the 5′ or 3′ copy of this sequence was selectively inactivated has recently shown that only the 5′ copy is active in packaging: deletion of the 3′ copy allows packaging, while deletion of the 5′ copy inactivates pregenome encapsidation (R. HIRSCH and D. GANEM, submitted). This is in contrast to the nearby sequences that specify the initiation of minus strand DNA synthesis, which (in WHV, at least) are only active in the 3′ position (SEEGER and MARAGOS 1990).

The location of the sequences that mediate RNA packaging does not explain the discrimination of the packaging machinary against the longer species of genomic RNAs, since the packaging signal is present in all three species. As noted previously, some other feature is required to explain this discrimination. Early studies (ENDERS et al. 1987) raised several possible models: e.g., that a secondary structure unique to pregenomic RNA allowed access to the packaging signal; that a specific relationship to the 5′ cap was necessary; or that translation of the preC region, which is only allowed by the longer transcripts containing the preC AUG, might sequester them from the packaging apparatus. Recently, NASSAL et al. (1990) have presented strong evidence in favor of the translation model, by showing that encapsidation of longer genomic RNAs occurs when the potential for preC translation is removed by mutation.

3 Subviral HBsAg Particle Formation

As noted previously, a hallmark of HBV infection is the overproduction of subviral particles composed only of the viral surface proteins and lipid. Although early immunologic studies suggested that only one viral antigen was present on these particles (and on virions), we now know that the repertoire of envelope proteins produced by the virus is more complex. This complexity, in turn, provides the diversity of structural elements that is required to build different structures out of shared components.

3.1 The Surface Antigen Family

The predominant viral surface protein in the circulation of infected individuals is the 24-kDa S protein, which (together with its glycosylated form) is present in concentrations of up to 50–300 μg/ml. The coding region for this protein was identified by aligning the translated HBV DNA sequence with the known N-terminal amino acid sequence of the S chain (VALENZUELA et al. 1979). This immediately revealed that the open reading frame for the S antigen was open for some 400 nt 5′ to the S AUG, and that this upstream (or preS) ORF could be divided into two subregions (preS1 and preS2) by the presence of two in-frame initiation codons (GALIBERT et al. 1979; VALENZUELA et al. 1980; see Fig. 1). Thus, the sequence raised the possibility that two larger, S-related proteins could be generated during infection by translation of this preS region. The preservation of this coding scheme in all the mammalian hepadnaviruses further supported the potential functional importance of this arrangement (GALIBERT et al. 1982; SEEGER et al. 1984). It was not long before both polypeptides were detected in all three mammalian hepadnaviruses (HEERMAN et al. 1984; WONG et al. 1985;

PERSING et al. 1986). The avian hepadnaviruses also contain a preS region with multiple in-frame AUGs, but only one preS protein has thus far been reproducibly detected (PUGH et al. 1987; SCHLICHT et al. 1987a).

In HBV, the largest of these so-called preS proteins is the 39-kDa L protein, which is the product of initiation at the first AUG of the ORF (HEERMAN et al. 1984). Initiation at the second AUG generates the 31-kDa M protein (MACHIDA et al. 1983, 1984; STIBBE and GERLICH 1983; PERSING et al. 1985). Both proteins share the common C-terminal S domain and differ principally by the length and structure of their N-terminal extensions. Both are quantitatively minor components of the circulating pool of S-related antigens, with the M protein accounting for approximately 5%–15% of the total, and the L protein typically representing only 1%–2% (or less) of the total. Each protein exists in two isomeric forms differing only by the presence or absence of S domain glycosylation; in addition, the M protein contains another N-linked oligosaccharide on its preS2-specific domain. The L protein contains an additional post-translational modification, a myristic acid group in amide linkage to its amino terminal glycine residue (PERSING et al. 1987).

As in the C coding region, the problem of ribosomal access to the internal AUGs is solved by the deployment of separate transcripts immediately 5′ to each AUG. In this case two promoters are required: the upstream promoter (P1) specifying the L mRNA, and the downstream (P2) specifying the M and S mRNAs (WILL et al. 1987; YAGINUMA et al. 1987). As might be expected from this arrangement, no precursor-product relationship exists between the preS and S proteins: frameshift or deletion mutations within preS do not affect S biosynthesis (PERSING et al. 1985). The relative levels of these transcripts in infected cells probably accounts for most of the differences in the stoichiometry of the circulating proteins. The L mRNA is present in very low quantities in infected liver (WILL et al. 1987), corresponding to its low concentration in serum, while S message is highly abundant. The case of the M protein requires further examination. Nuclease mapping studies show that M message can be present at levels of up to 30%–50% that of S mRNA (CATTANEO et al. 1983; STANDRING et al. 1984), yet the relative levels of M to S protein are usually measurably less than this. Since the stabilities of these two proteins are not markedly different, this may indicate a modest role for translational control in M protein expression.

3.2 The S Protein and the Assembly of 20-nm Particles

The 20 nm particles purified from carrier serum typically contain approximately 100 polypeptide subunits per particle, predominantly the S protein with variable amounts of M and only trace quantities of L subunits (HEERMAN et al. 1984). With the advent of cloned HBV DNA, it became possible to express the surface protein coding regions in heterologous cells, leading to the discovery that cells expressing M and S subunits assemble and secrete 20-nm particles that are indistinguishable from those found in serum (DUBOIS et al. 1980). This observ-

Mixed M/S particles are efficiently secreted by most cells (MICHEL et al. 1984; PERSING et al. 1985; CHENG and MOSS 1987). Studies reporting that M protein overexpression is inhibitory to S export (STANDRING et al. 1986) were conducted in *Xenopus* oocytes, where the efficiency of S export is already very low; inhibitory effects do not occur in mammalian cells. Particles containing both M and S subunits are of great interest as potential vaccines because of their enhanced immunogenicity, as documented in immunization studies of inbred mice. Even mice genetically incapable of responding to S determinants will make an anti-S response following immunization with such particles (MILICH et al. 1985). It is hoped that similar results can be achieved in the appreciable number of humans who respond poorly or not at all to S protein-based vaccines (ALPER et al. 1989).

Recent studies have sought to define subdomains of the S protein that might mediate one or more of the steps in S particle assembly (V. BRUSS and D. GANEM, in preparation). While simple in principle, such experiments are bedeviled in practice by the unpredictable effects of mutations (especially deletions) on protein folding and stability. Some conclusions can nonetheless be made. First, insertions into one region of the cytoplasmically disposed loop region between signals I and II are generally well tolerated, with minimal effects on particle assembly and secretion (DELPEYROUX et al. 1986, 1987; V. BRUSS and D. GANEM, unpublished results). Lesions affecting the proper disposition of the N-terminus (e.g., deletions of signal I, substitution of cleaved signals for signal I, or addition of bulky extensions to the N-terminus) are poorly tolerated; although they enter the secretory pathway (as judged by glycosylation), they fail to assemble either with each other, or with wild-type S chains when these are coexpressed with them. Interestingly, the ability of one such molecule (a globin–S fusion protein in which approximately 100 aa of alpha globin were added to the N-terminus of S) to be efficiently assembled and secreted was rescued by restoring its correct transmembrane topology through provision of an N-terminal signal sequence (V. BRUSS and D. GANEM, in preparation). These data attest to the importance of the proper orientation of the first portions of the S chain in allowing lateral interactions with other monomers. Consistent with this, an amber mutant missing the C-terminal hydrophobic domain of S remains competent to aggregate with other S chains, since it is efficiently cosecreted with wild-type S chain when the two are coexpressed. Clearly, further understanding of the details of S particle assembly will require both more refined genetic studies and the development of an in vitro system in which correct particle formation can be observed. One useful byproduct of such an understanding could be the ability to rationally design chimeric particles containing foreign epitopes for possible use as vaccines. Several successful chimeras have already been generated empirically (DELPEYROUX et al. 1986; 1987; VALENZUELA et al. 1985; BRUSS and GANEM 1991), but progress has been slow due to both the lack of precise information about S assembly domains and to the twin problems of misfolding and instability mentioned above.

3.3 The L Protein and 20-nm Particle Formation

We have previously noted that natural 20-nm spherical particles often lack detectable L protein, and that the L protein is overall a minor component of the circulating pool of surface antigens (HEERMAN et al. 1984). The reason for the low abundance of L chains on 20-nm particles became clearer in experiments aimed at overexpression of ths subunit in cultured cells. Vaccinia vectors that express only the L protein revealed that the L chains themselves are not secreted, despite the presence of an intact S domain that contains all the information required for this process (CHENG et al. 1986). This suggests that the preS1-specific domain of L contains signals that promote intracellular retention of the chain and that can override the secretory elements in the S domain. While the topology of this protein was not examined in these experiments, in other studies in *Xenopus* oocytes we have observed that the protein remains in a transmembrane configuration and does not progress to the later, protease-resistant forms that characterize normal S morphogenesis (K. SIMON and D. GANEM, unpublished results). Epitope mapping studies with preS1-specific monoclonal antibodies suggest that most of the preS1 domain is disposed on the lumenal side of the ER membrane (KUROKI et al. 1990).

When L chains are overexpressed in the presence of M and S chains, an additional finding emerges: that the ability of the S and M chains to be secreted is specifically inhibited in a dose-dependent fashion (CHENG et al. 1986; PERSING et al. 1986; STANDRING et al. 1986; CHISARI et al. 1986). All three chains accumulate intracellularly, and examination of their carbohydrate side chains indicates that they are retained in a pre-Golgi compartment; electron-microscopic studies confirm this to be the endoplasmic reticulum (CHISARI et al. 1987). Presumably this retention reflects the formation of mixed arrays of L-M-S aggregates, with the inhibitory influence of the preS1 domain now being conferred upon the entire aggregate. The inhibitory effect of L overexpression on S particle secretion explains the underrepresentation of L chains on 20-nm spherical particles. Interestingly, in electron-microscopic studies of the hepatocytes of transgenic mice overexpressing L chains (relative to S), 20-nm filamentous forms were frequently observed in the ER (CHISARI et al. 1987). This accords well with earlier observations that filaments are relatively enriched in L subunits (HEERMAN et al. 1984) and suggests that filaments arise in regions of the ER where local concentrations of L protein are high.

The overaccumulation of retained L-M-S aggregates can be injurious to the host cell. Hepatocytes of transgenic mice bearing these aggregates develop characteristic cytopathic effects that over long periods of time provoke hepatocellular regeneration and ultimately hepatocellular carcinoma (CHISARI et al. 1987, 1989). The implications of these findings for viral pathogenesis will be considered elsewhere in this volume. Suffice it to say here that in liver biopsy specimens from cases of acute hepatitis B cells with similar cytopathic effects (so-called ground glass hepatocytes) are not infrequently seen, and it seems likely that these reflect the consequences of L protein overproduction in these cells.

What accounts for the intracellular retention of surface antigens by L chains? Clearly this property is specified, directly or indirectly, by the preS1 domain. Deletion mutagenesis of this region indicates that the inhibitory sequences map to the extreme N-terminus of the molecule (KUROKI et al. 1989). The observation that this region of the L protein is myristylated raised the possibility that ER retention might be mediated by the affinity of this hydrophobic fatty acid for the ER membrane. In other cellular proteins known to harbor this modification it is frequently (but not invariably) involved in promoting associations with cellular membranes (TOWLER et al. 1988). Examinations of the role L protein myristylation in secretion have led to divergent results. Missense mutations that ablate the ability of the L protein of the *adw* serotype of HBV to be myristylated clearly have no effect on secretion-inhibition, indicating that in this serotype primary sequence features play the dominant role (KUROKI et al. 1989). However, similar studies employing the *ayw* serotype suggest a prominent role for myristylation in this activity (R. PRANGE, A. CLEMEN, and R. STREECK, personal communication). These differences are most likely attributable to the fact that the sequences of these two L proteins differ significantly at their N termini, with the *adw* serotype containing an insertion of an added 11 amino acids in this position. This molecule is more potent as an inhibitor of secretion than is its *ayw* counterpart (R. RUSSNAK and D. GANEM, unpublished results), though both are functional in vivo, as judged by the infectivity of their parent clones.

4 Dane Particle Assembly

The extensive progress of the past few years in understanding the biogenesis of subviral HBsAg particles has provided a firmer basis for thinking about the mechanism of Dane particle assembly. Since all three envelope proteins are expressed primarily on intracellular membranes rather than on the cell surface (OU and RUTTER 1987), virion formation almost certainly proceeds on intracellular membranes; electron-microscopic studies consistent with this have been reported (KAMIMURA et al. 1981). Perhaps the most important additional advance in this area was the recognition by Gerlich and colleagues that the envelope polypeptide composition of virions differs significantly from that of 20-nm spheres (HEERMAN et al. 1984). They showed, that unlike the latter, virions contained relatively large quantities of L proteins; in their preparations the ratio of L:M:S subunits was approximately 1:1:4. Since these preparations are unavoidably contaminated to some extent with 20-nm particles, it is likely that the ratio of L:S chains in authentic virions is still higher.

More recent studies have used genetic approaches to define the components required for Dane particle production. Cultured hepatoma cells (HepG2, Huh6, and Huh7) transfected with HBV DNA secrete infectious virions (SUREAU et al. 1986; TSURIMOTO et al. 1987; YAGINUMA et al. 1987; SELLS et al. 1987). By transfecting these cells with missense or amber mutants selectively affecting the L, M, or S proteins, two groups have shown that the L and S proteins are required for correct virion morphogenesis (K. UEDA, T. TSURIMOTO, and K. MATSUBARA, unpublished results; BRUSS and GANEM 1991). (The role of the M protein in assembly is still disputed.) Unlike retroviruses, whose nucleocapsids can still bud out of cells in the absence of envelope proteins, no HBV cores are exported if either of these proteins is absent. HBV missense mutants affecting L myristylation remain competent for virion assembly and secretion (BRUSS and GANEM 1991); although the infectivity of these mutants is difficult to test, analogous mutants of DHBV are known to be noninfectious in the animal host (MACRAE, BRUSS and GANEM 1991). If, as seems likely, the same is true for HBV, then myristylation may play its most important role in the adsorption, entry, or uncoating of virions on the next cycle of infection.

The secretary behavior of the L protein and its enrichment on virions have led to a model for virion assembly (BRUSS and GANEM 1991) that attributes a central role in this process to the L chain. In this view, L is the analog of the conventional envelope glycoproteins of other viruses (e.g., VSV G or influenza HA), moored in the membrane awaiting interactions with nucleocapsids. In this view, the interaction of such chains with cores would precipitate the same series of morphogenetic events that for the S proteins occurs constitutively, perhaps by triggering a conformational change in the L polypeptide. The genetic data indicate that the S proteins must also play an active role in the budding process, since L proteins cannot perform this act alone. Perhaps the precise stoichiometry of the L and S chains is important; suboptimal ratios may lead instead to filament formation.

Clearly, much remains to be explained. How do cores trigger budding? Which regions of core antigen interact with which regions of the surface protein(s)? Why do only cores containing mature (replicated) genomes appear extracellularly? Do additional morphologic changes occur distal to the budding step, as virions are transported through the secretory pathway? These and other lingering questions will surely preoccupy students of hepadnaviral morphogenesis for many years to come.

References

Aggerbeck LP, Peterson DL (1985) Electron microscopic and solution X-ray scattering observations on the structure of hepatitis B surface antigen. Virology 141: 155–161

Albin C, Robinson WS (1980) Protein kinase activity in hepatitis B virions. J Virol 34: 297–302

Alper CA, Kruskall MS, Marcus-Bagley D, Craven DE, Katz AJ, Brink SJ, Dienstag JL, Swdeh Z, Yunis EJ (1989) Genetic prediction of nonresponse to hepatitis B vaccine, N Engl J Med 321: 708–712

Bartenschlager R, Schaller H (1988) The amino terminal domain of the hepadnaviral P gene encodes the terminal protein (genome-linked protein) believed to prime reverse transcription. EMBO J: 4185–4192

Bayer M, Blumberg B. Werner B (1968) Particles associated with Australia antigen in the sera of patients with leukemia, Down's Syndrome and hepatitis. Nature 218: 1057–1059

Birnbaum F and Nassal M (1990) Hepatitis B virus nucleocapsid assembly: Primary structure requirements in the core protein J Virol 64: 3319–3330

Bartenschlager R, Junker-Niepmann M and Schauer H (1990) The P protein of HBV is required as a structural component for genomic RNA encapsidation J Virol 64: 5324–5332

Blumberg BS, Alter HJ, Visnich (1965) A "new" antigen in leukemia sera. JAMA 191: 541–546

Brunetto M, Stemmler M, Schodel F, Will H, Ottobrelli A, Rizetto M, Verme, Bonino F (1989) Identification of HBV variants which cannot produce precore-derived HBeAg and may be responsible for severe hepatitis. Ital J Gastroenterol 21: 151–154

Bruss V and Ganem D (1991) The role of envelope proteins in hepatitis B virus assembly. Proc Nat Acd Sci USA 88: 1059–1063

Bruss V, Gerlich WH (1988) Formation of transmembranes hepatitis B e-antigen by cotranslational *in vitro* processing of the viral precore protein. Virology 163: 268–275

Burrell CJ, Mackay P, Greenway PJ, Hofschneider PH, Murray K (1979) Expression in *Escherichia coli* of hepatitis B virus DNA sequences cloned in plasmid pBR322. Nature 279: 43–47

Büscher M, Reiser W, Will H, Schaller H (1985) Transcripts and the putative RNA pregenome of duck hepatitis B virus: implications for reverse transcription. Cell 40: 717–724

Cattaneo R, Will H, Hernandez N, Schaller H (1983) Signals regulating hepatitis B surface antigen transcription. Nature 305: 336–338

Chang C, Enders G, Sprengel R, Peters N, Varmus HE, Ganem D (1987) Expression of the precore region of an avian hepatitis B virus is not required for viral replication. J Virol 61: 3322–3325

Cheng C, Smith K, Moss B (1986) Hepatitis large surface protein is not secreted but is immunogenic when selectively expressed by recombinant vaccinia virus. J Virol 60: 337–344

Cheng KC, Moss B (1987) Selective synthesis and secretion of particles composed of the hepatitis B virus middle surface protein directed by a recombinant vaccinia virus: induction of antibodies to preS and S epitopes. J Virol 61: 1286–1290

Chisari FV, Fillipi P, MacLachlan A, Milich D, Riggs M, Lee S, Palmiter R, Pinkert C, Brinster R (1986) Expression of hepatitis B large envelope polypeptide inhibits hepatitis B surface antigen secretion in transgenic mice. J Virol 60: 880–887

Chisari FV, Filippi P, Buras J, MacLachlan A, Popper H, Pinkert C, Palmiter R, Brinster R (1987) Structural and pathological effects of synthesis of hepatitis B virus large envelope polypeptide in transgenic mice. Proc Natl Acad Sci USA 84: 6909–6913

Chrisari FV, Klopchin K, Moriyama T, Pasquinelli C, Dunsford HA, Brinster RL, Palmiter RD (1989) Molecular pathogenesis of hepatocellular carcinoma in hepatitis B virus transgenic mice. Cell 59: 1145–1156

Clarke BE, Newton SE, Carroll AR, Francis MJ, Appleyard G, Syred AD, Highfield PE, Rowlands DJ, Brown F (1987) Improved immunogenicity of a peptide epitope after fusion to hepatitis B core protein. Nature 330: 381–384

Cohen B, Richmond J (1982) Electron microscopy of hepatitis B core antigen synthesized in *C. coli*. Nature 296: 677–678

Dane DS, Cameron CH, Briggs M (1970) Virus-like particles in serum patients with Australia antigen-associated hepatitis. Lancet 1: 695–698

Delpeyroux F, Chenciner N, Lim A, Malpiece Y, Blondel B, Crainic R, van der Werf, Streeck RE (1986) A poliovirus neutralization epitope expressed on hybrid hepatitis B surface antigen particles. Science 233: 472–475

Delpeyroux F, Chenciner N, Lim A, Lambert M, Malpiece Y, Streeck RE (1987) Insertions in the hepatitis B surface antigen: effect on assembly and secretion of 22-nm particles from mammalian cells. J Mol Biol 195: 343–350

Delpeyroux F, Peillon N, Blondel B, Crainic R, Streeck RE (1988) Presentation and immunogenicity of the hepatitis B surface antigen and poliovirus neutralization antigen on mixed empty envelope particles. J Virol 62: 1836–1839

Dubois MF, Pourcel C, Rousset S, Chany C, Tiollais P (1980) Excretion of hepatitis B surface antigen particles from mouse cells transformed with cloned viral DNA. Proc Natl Acad Sci USA 77: 4549–4553

Eble BE, Lingappa VR, Ganem D (1986) Hepatitis B surface antigen: an unusual secreted protein initially synthesized as a transmembrane polypeptide. Mol Cell Biol 6: 1454–1463

Eble BE, Macrae DR, Lingappa VR, Ganem D (1987) Multiple topogenic sequences determine the transmembrane orientation hepatitis B surface antigen. Mol Cell Biol 7: 3591–3601

Eble BE, Lingappa VR, Ganem D (1990) The N-terminal (preS2) domain of a hepatitis B virus surface glycoprotein is translocated across membranes by downstream signal sequences. J Virol 64: 1414–1419

Enders GH, Ganem D, Varmus HE (1985) Mapping the major transcripts of ground squirrel hepatitis virus: the presumptive template for reverse transcriptase is terminally redundant. Cell 42: 297–308

Enders GH, Ganem D, Varmus HE (1987) 5′-terminal sequences influence the segregation of ground squirrel hepatitis virus RNAs into polyribosomes and viral core particles. J Virol 61: 35–41

Feitelson M, Marion P, Robinson W (1982) Core particles of hepatitis B virus and ground squirrel hepatitis virus II: characterization of the protein kinase reaction associated with ground squirrel hepatitis virus and hepatitis B virus. J Virol 43: 741–748

Galibert F, Mandart E, Fitoussi F, Tiollais P, Charnay P (1979) Nucleotide sequence of hepatitis B virus genome (subtype ayw) cloned in *E. coli*. Nature 281: 646–650

Galibert F, Chen TN, Mandart E (1982) Nucleotide sequence of a cloned woodchuck hepatitis virus genome: comparison with the hepatitis B virus sequence. J Virol 41: 51–65

Gallina A, Bonelli F, Zentilin L, Rindi G, Muttini M, Milanesi G (1989) A recombinant hepatitis B core antigen polypeptide with the protamine-like domain deleted self-assembles into capsid particles but fails to bind nucleic acids. J Virol 63: 4645–4652

Ganem D (1982) Persistent infection of humans with hepatitis B virus: mechanisms and consequences. Rev Infect Dis 4: 1026–1047

Ganem D, Varmus HE (1987) The molecular biology of the hepatitis B viruses. Annu Rev Biochem 56: 651–694

Garcia P, Ou J, Rutter WJ, Walter P (1988) Targeting of protein of hepatitis B virus to the endoplasmic reticulum membrane: after signal peptide cleavage translocation can be aborted and the product released into the cytoplasm. J Cell Biol 106: 1093–1104

Gavilanes F, Gonzales-Ros A, Peterson D (1982) Structure of hepatitis B surface antigen: characterization of the lipid components and their association with the viral proteins. J Biol Chem 257: 7770–7777

Gerber M, Hadziyannis S, Vissoulis C, Schaffner F, Paronetto F, Popper H (1974) Electron microscopy and immunoelectron microscopy of cytoplasmic hepatitis B antigen in hepatocyte. Am J Pathol 75: 489–502

Gerin J, Purcell R, Hoggan M, Holland P, Chanock R (1969) Biophysical properties of Australian antigen. J Virol 36: 787–795

Gerlich W, Goldmann U, Muller R, Stibbe W, Wolff W (1982) Specificity and localization of the hepatitis B virus-associated protein kinase. J Virol 42: 761–766

Gocke D (1975) Extrahepatic manifestation of viral hepatitis. Amer J Med Sci 207: 49–55

Heermann KH, Goldmann I, Schwartz W, Seyffarth T, Baumgarten H, Gerlich WH (1984) Large surface proteins of hepatitis B virus containing the preS sequence. J Virol 52: 396–402

Hemmingsen SM, Woolford C, van der Vies SM, Tilly K, Dennis DT, Georgopoulos CP, Hendrix RW, Ellis RJ (1988) Homologous plant and bacterial proteins chaperone oligomeric protein assembly. Nature 333: 330–334

Hirsch R, Lavine J, Chang L, Varmus H, Ganem D (1990) Polymerase gene products of hepatitis B viruses are required for genomic RNA packaging as well as for reverse transcription. Nature 344: 522–555

Hoofnagle JH, Gerety RJ, Barker LF (1975) Antibody to hepatitis B core antigen. Am J Med Sci 270: 179–187

Hoofnagle JH, Michalak T, Nowoslawski A, Gerety RJ, Barker (1978) Immunofluorescence microscopy in experimentally induced type B hepatitis in the chimpanzee. Gastroenterology 74: 182–187

Jean-Jean O, Levrero M, Will H, Perricaudet M, Rossignol JM (1989a) Expression mechanism of the hepatitis B virus (HBV) C gene and biosynthesis of HBe antigen. Virology 170: 99–106

Jean-Jean O, Salhi S, Carlier D, Elie C, DeRecondo A-M, Rossignol J-M (1989b) Biosynthesis of hepatitis B virus e antigen: directed mutagenesis of the putative aspartyl protease site. J Virol 63: 5497–5500

Junker-Niepmann M, Bartenschlager R and Schaller H (1990) A short *cis*-acting sequence is required for hepatitis B virus pregenome encapsidation and sufficient for packaging of foreign RNA EMBO 9: 3389–3396

Kamimura T, Yoshikawa A, Ichida F, Sasaki H (1981) Electron microscopy studies of Dane particles in

Schlicht HJ, Bartenschlager R, Schaller H (1989) The duck hepatitis B virus core protein contains a highly phosphorylated C-terminus that is essential for replication but not for RNA packaging. J Virol 63: 2995–3000
Seeger C, Maragos J (1990) Identification and characterization of the woodchuck hepatitis virus origin of DNA replication. J Virol 64: 16–23
Seeger C, Ganem D, Varmus HE (1984) Nucleotide sequence of an infectious molecularly cloned genome of ground squirrel hepatitis virus. J Virol 51: 367–375
Seeger C, Ganem D, Varmus HE (1986) Genetic and biochemical evidence for the hepatitis B virus replication strategy. Science 232: 477–485
Sells MA, Chen ML, Acs G (1987) production of hepatitis B virus particles in Hep G2 cells transfected with cloned hepatitis B virus DNA. Proc Natl Acad Sci USA 84: 1005–1009
Shaeffer E, Snyder R, Sninsky J (1986) Identification and localization of preS-encoded polypeptides from woodchuck and ground squirrel hepatitis viruses. J Virol 57: 173–182
Shih J, Gerin J (1977) Proteins of hepatitis B surface antigen: amino acid compositions of the major polypeptides. J Virol 21: 1219–1222
Simon K, Lingappa V, Ganem D (1988a) Secreted hepatitis B surface antigen polypeptides are derived from a transmembrane precursor. J Cell Biol 107: 2163–2168
Simon K, Lingappa V, Ganem D (1988b) A block to the intracellular transport and assembly of hepatitis B surface antigen polypeptides in Xenopus oocytes. Virology 166: 76–81
Speiss M, Lodish H (1986) An internal signal sequence: the asialoglycoprotein membrane anchor. Cell 44: 177–185
Stah SJ, Murray K (1989) Immunogenicity of peptide fusions to hepatitis B virus core antigen. Proc Natl Acad Sci USA 86: 6283–6287
Standring DN, Rutter WJ, Varmus HE, Ganem D (1984) Transcription of the hepatitis B surface antigen gene in cultured murine cells initiates within the presurface region. J Virol 50: 563–571
Standring DN, Ou J, Rutter WJ (1986) Assembly of viral particles in Xenopus oocytes: pre-surface-antigens regulate secretion of the hepatitis B viral surface envelope particle. Proc Natl Acad Sci USA 83: 9338–9342
Standring DN, Ou J, Masiarz F, Rutter WJ (1988) A signal peptide encoded within the precore region of hepatitis B virus directs the secretion of a heterogeneous population of e antigens in Xenopus oocytes. Proc Natl Acad Sci USA 85: 8405–8409
Stenlund A, Lamy D, Moreno-Lopez J, Ahola H, Pettersson U, Tiollais P (1983) Secretion of the hepatitis B virus surface antigen from mouse cells using an extra-chromosomal eukaryotic vector. EMBO J 2: 669–673
Stibbe W, Gerlich W (1983) Structural relationships between minor and major proteins of hepatitis B surface antigen. J Virol 46: 626–628
Stratowa C, Doehmer J, Wang Y, Hofschneider PH (1982) Recombinant retroviral DNA yielding high expression of hepatitis B surface antigen. EMBO J 1: 1573–1578
Summers J, Mason WS (1982) Replication of the genome of a hepatitis B-like virus by reverse transcription of an RNA intermediate. Cell 29: 403–415
Summers J, O'Connell A, Millman I (1975) Genome of hepatitis B virus: restriction enzyme cleavage and structure of DNA extracted from Dane particles. Proc Natl Acad Sci USA 72: 4597–4601
Sureau C, Romet-Lemonne JL, Mullins JI, Essex M (1986) Production of hepatitis B virus by a differentiated human hepatoma cell line after transfection with cloned circular HBV DNA. Cell 47: 37–47
Szmuness W, Stevens CE, Harley EJ, Zang EA, Oleszko WR, William DC, Sadovsky R, Morrison JM (1980) Hepatitis B vaccine. N. Engl J Med 303: 833–841
Towler D, Gordon J, Adams S, Glaser L (1988) The biology and enzymology of eukaryotic protein acylation. Annu Rev Biochem 57: 69–99
Townsend A, Bodmer H (1989) Antigen recognition by class I-restricted T lymphocytes. Annu Rev Immunol 7: 601–624
Tsurimoto T, Fujiyama A, Matsubara K (1987) Stable expression and replication of hepatitis B virus genome in an integrated state in a human hepatoma cell line transfected with the cloned viral DNA. Proc Natl Acad Sci USA 84: 444–448
Tuttleman J, Pourcel C, Summers J (1986) Formation of the pool of covalently closed circular viral DNA in hepadnavirus-infected cells. Cell 47: 451–460
Uy A, Bruss V, Gerlich W, Kochel H, Thomson R (1986) Precore sequence of hepatitis B virus including a antigen and membrane association of the viral core protein. Virology 155: 89–96
Valenzuela P, Gray P, Quiroga M, Zaldivar J, Goodman HM, Rutter WJ (1979) Nucleotide sequence of the gene coding for the major protein of hepatitis B virus surface antigen. Nature 280: 815–819

Valenzuela P, Quiroga M, Zaldivar J, Gray P, Rutter WJ (1980) The nucleotide sequence of the hepatitis B viral genome and the identification of the major viral genes. In: Fields BN, Jaenisch R (eds) Animal virus genetics ICN/UCLA Symp Mol Cell Biol, pp 57–70

Valenzuela P, Medina A, Rutter WJ (1982) Synthesis and assembly of hepatitis B virus surface antigen particles in yeast. Nature 298: 347–351

Valenzuela P, Coit D, Medina-Selby A, Kuo C, Van Nest G, Burke RL, Bull P, Urdea M, Graves P (1985) Antigen engineering in yeast: Synthesis and assembly of hybrid hepatitis B surface antigen—herpes simply 1 gD particles. Biotechnology 3: 323–326

Wampler DE, Lehman ED, Boger J, McAleer WJ, Scolnick EM (1985) Multiple chemical forms of hepatitis B surface antigen produced in yeast. Proc Natl Acad Sci USA 82: 6830–6834

Weimer T, Salfeld J, Will H (1987) Expression of hepatitis B virus core gene in vitro and in vivo. J Virol 61: 3109–3113

Will H, Cattaneo R, Pfaff E, Kuhn C, Roggendorf M, Schaller H (1984) Expression of hepatitis B antigen with a simian virus 40 infection. J Virol 50: 335–342

Will H, Reiser W, Weimer T, Pfaff E, Buscher M, Sprengel R, Cattaneo R, Schaller H (1987) Replication strategy of human hepatitis B virus. J Virol 61: 904–911

Wong DT, Nath N, Sninsky JJ (1985) Identification of hepatitis B virus polypeptides encoded by the enitre preS open reading frame. J Virol 55: 223–231

Yaginuma K, Shirakata Y, Kobayashi M, Koike K (1987) Hepatitis B virus (HBV) particles are produced in a cell culture system by transient expression of transfected BHV DNA. Proc Natl Acad Sci USA 84: 2678–2682

Yamada G, Feinberg L, Nakane P (1978) Hepatitis B: cytologic localization of virus antigens and the role of the immune response. Human Path 9: 93–109

Yeh C-T, Liaw Y-F and Ou J-H (1990) The arginine-rich domain of HBV precore and core proteins contains a signal for nuclear transport. J virol 64: 6141–6147

During the past few years, several laboratories have taken advantage of this opportunity to expand our understanding of the biology of this virus and to elucidate some of the reciprocal interactions of virus and host that may determine the outcome of infection. In this chapter I will review this new information and discuss its implications in terms of the biology and pathogenesis of HBV. Since the model system entails the introduction of the HBV genome or its constituent open reading frames (ORFs) into transgenic animals, a brief overview of selected aspects of the viral genome will be presented before describing the studies themselves. This background material is reviewed in greater detail in other chapters of this volume.

2 Products of the HBV Genome

The HBV virion contains a circular, partially double-stranded DNA genome approximately 3200 base pairs in length (Fig. 1). The long (minus) strand encodes the viral mRNA and the viral pregenomic RNA which is reverse transcribed as an early step in viral replication. The minus strand contains four long, overlapping ORFs that encode structural (envelope, nucleocapsid) and nonstructural (polymerase, transactivating protein) viral elements.

The nucleocapsid ORF contains two in-phase start codons that define two overlapping polypeptides, the shorter of which (core, HBcAg) is a cytoplasmic and nuclear protein which represents the viral nucleocapsid. Recent studies indicate that HBcAg is expressed on the surface of circulating infectious virus particles (MOLLER et al. 1989). Its longer counterpart (precore) is a secreted, cytoplasmic, nuclear, and membrane protein (OU et al. 1989; SCHLICHT and SCHALLER 1989), whose function is currently undefined. Based on considerable circumstantial clinical evidence, it is believed that a cellular immune response to the nucleocapsid antigens is an important factor in viral clearance in HBV infection (reviewed in CHISARI et al. 1989b).

The envelope ORF contains three in-phase translation start codons which define the amino termini of three overlapping polypeptides, whose expression is transcriptionally regulated (see chapter by H. SCHALLER, this volume). All three are components of the virus envelope. In addition, the major and middle envelope polypeptides assemble into small, 22-nm spherical particles by budding into the lumen of the endoplasmic reticulum (ER) and they are rapidly secreted after passage and glycosylation in the Golgi apparatus (PATZER et al. 1984, 1986; EBLE et al. 1986, 1987). The large envelope polypeptide (L) also exerts an important structural influence on the formation of the abundant, noninfectious, subviral particles that are characteristic of HBV infection (CHISARI et al. 1986; STANDRING et al. 1986; MCLACHLAN et al. 1987; OU and RUTTER 1987; MOLNAR-KIMBER et al. 1988). Depending on the relative molar ratio of the large envelope polypeptide to the major and middle envelope polypeptides within a

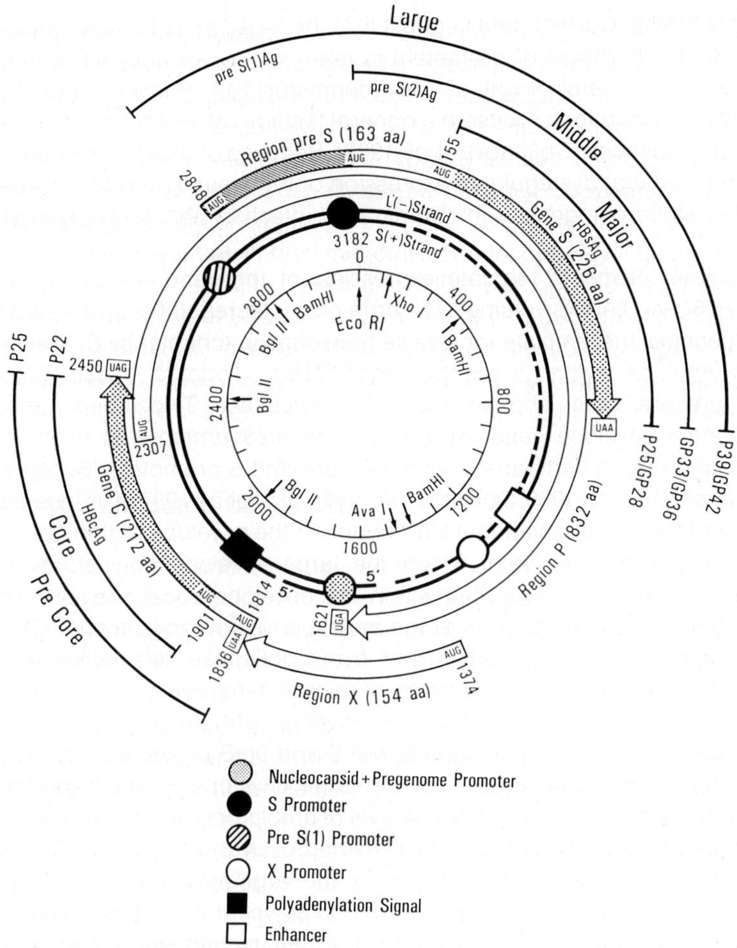

Fig. 1. Map of hepatitis B virus genome

given cell, either short secretable filaments or long, nonsecretable, branching filaments that accumulate within the ER are produced. Intracellular accumulation of these nonsecretable filaments is cytopathic, producing so-called ground glass hepatocytes when this process occurs in infected liver cells. In fact, this process may ultimately culminate in the death of the cell (CHISARI et al. 1987). Recent evidence (MORIYAMA et al. 1990) indicates that the envelope-specific, cytotoxic T cells can also cause hepatocellular injury, at least in transgenic mice, suggesting that these envelope polypeptides may also be important targets for immune-mediated liver disease during HBV infection in humans.

The X ORF encodes a transcriptional *trans*-activating protein that positively regulates transcription from HBV and other viral and cellular promoters (TWU and SCHLOEMER 1987; SPANDAU and LEE 1988; SETO et al. 1988; TWU et al. 1989;

of the existing transgenic lineages despite evidence of viral replication and expression in lymphoid tissues in infected humans and woodchucks (KORBA et al. 1989). Other than obvious species differences, the reason for this inconsistency is not known.

The fairly widespread expression of the transgenic HBV genome in many tissues together with mounting evidence of extrahepatic hepadnavirus expression during natural infection (KORBA et al. 1986) suggests that the relative tissue and species specificity of HBV is a reflection of multiple constraints at the levels of viral entry, replication, and gene expression, and that none of these constraints is absolutely specific for the human hepatocyte. It also appears that the HBV genome contains *cis*-acting regulatory elements that are preferentially responsive to tissue-specific transacting factors that are particularly abundant in the liver and kidney.

3.3 Viral Replication in Transgenic Mice

Following microinjection of greater-than-genome length constructs capable of encoding all of the viral transcripts as well as the pregenomic viral RNA, several groups have demonstrated the presence of small amounts of cytoplasmic single-stranded and double-stranded HBV DNA replicative intermediates in liver, kidney, and heart and complete virus-like particles in the serum (FARZA et al. 1988; ARAKI et al. 1989b; CHISARI et al., unpublished data), indicative of low level viral replication. During natural infection, the cytoplasmic replicative intermediates are transported within nascent core particles to the nucleus where they mature into covalently closed circular (supercoiled) viral DNA molecules that serve to amplify the transcriptional template for viral replication (chapter by C. SEEGER et al., this volume). For unknown reasons, this process is inefficient in transgenic mice, and transcription is limited to the small number of integrated transgenes present in each cell. The absence of intranuclear, supercoiled, viral genomes may also contribute to the apparent absence of secondary viral integrations in these lineages.

The 2.1-kb (HBs) transcript is most abundant and widespread in its tissue distribution in these lineages. The 3.5-kb (RNA pregenome/nucleocapsid) transcript (chapter by H. SCHALLER and M. FISCHER, this volume) is usually less abundant and is preferentially expressed in liver, kidney, and heart. Viral DNA replication is, indeed, limited to these three tissues, a finding consistent with the fact that DNA synthesis initiates with reverse transcription of pregenomic RNA. As in natural infection, the 2.4-kb (preS) and the 0.8-kb (X) transcripts are rare in transgenic mice. The preS transcript has been found only in the liver and kidney where it presumably contributes to the formation of complete virus particles. The X transcript is also present in liver and kidney where it presumably augments the expression of the remaining viral gene products. For unknown reasons, the X transcript is also commonly found in the testis where it is usually the only HBV-encoded transcript present.

3.4 Regulation of HBV Gene Expression in Transgenic Mice

Integration of HBV sequences is a natural consequence of chronic infections and is generally believed to play some role in the final outcome of the disease. As discussed in the chapter by C.E. ROGLER (this volume), integration may have a direct role in carcinogenesis. Expression from integrated DNA also explains the continued synthesis of HBsAg when all episomal forms of HBV DNA have disappeared. It is less clear why HBcAg is not expressed, and there is still uncertainty about the expression of *pol* and X-gene products from integrated DNA. The possible expression of X from integrated DNA is especially intriguing because of speculation about the role of this transactivator in the development of liver cancer. Unfortunately, analysis of these issues in infected livers presents practical and technical problems that are virtually insurmountable. Some of them are approachable in the transgenic mouse.

Analysis of assorted independent transgenic lineages demonstrates that HBV gene expression is regulated at multiple levels. Several groups have documented an association between the degree of methylation and transcription. Indeed, lack of HBV expression has been associated with hypermethylation of the transgene sequences in numerous transgenic lineages (YAMAMURA et al. 1987; HADCHOUEL et al. 1987; FARZA et al. 1988; POURCEL 1989; EL-GHOR and BURK 1989; ARAKI et al. 1989a). Further, HBV sequences are often hypomethylated in tissues that express the corresponding HBV genes in contrast to those that do not; these associations, however, are not absolute (POURCEL 1989; EL-GHOR and BURK 1989). Demethylation of HBV sequences has been associated with the onset of HBV gene expression during embryonic development (POURCEL 1989). Similarly, demethylation in vivo by treatment with 5-azacytidine is associated with expression of HBsAg in transgenic mice that carry heavily methylated, nonexpressed HBV sequences (ARAKI et al. 1989a). Finally, a transgenic lineage has been described in which HBV expression is irreversibly extinguished when the integrated HBV sequences are methylated de novo upon passage through the female germline (HADCHOUEL et al. 1987). Although this observation is unique to a single transgenic lineage and is, therefore, most probably a position effect, it further underscores the potential relevance of hypermethylation to suppression of HBV gene expression during the course of natural HBV infections, e.g., in the regulation of acute infections or in the shut-off of virus production late in the course of chronic infections.

As in other systems, hypomethylation may be necessary, but it clearly is not sufficient for the expression of integrated HBV DNA. In transgenic mice and during chronic HBV infection, expression of integrated viral DNA sequences probably depends upon the cooperative effects of multiple additional factors such as chromatin structure, cellular flanking sequences, and exposure to cellular- and virus-encoded DNA-binding proteins that regulate viral gene expression in addition to the methylation status of the integrated DNA. Along these lines, EL-GHOR and BURK (1989) have recently demonstrated a positive correlation between tissue-specific expression of HBV sequences and the presence of DNase I-hypersensitive sites in the HBV enhancer element and the core promoter region in chromatin obtained from different tissues of two transgenic mouse lineages. If the presence and locatiòn of DNase 1-hypersensitive sites reflects a change in chromatin structure secondary to local transcriptional activity, it is possible that they represent the site of interaction between regulatory regions of the viral genome and cellular transcription factors. Since several cellular transcription factors have been described in recent years that interact with HBV sequences in the enhancer and core promoter regions (chapter by H. SCHALLER, this volume), it is reasonable to suspect that similar interactions may contribute to the expression of the integrated viral genome in these transgenic mice.

The influence of flanking cellular sequences on transgene expression (position effect) is dramatically illustrated by the integration site-dependent maternal imprinting of HBV gene expression in a transgenic lineage described above (HADCHOUEL et al. 1987). It is further shown in the variable frequency and distribution of HBsAg-positive hepatocytes in 10 independent transgenic lineages containing the identical transgene at different integration sites (CHISARI et al. 1987). The percentage of HBsAg-positive hepatocytes in these different transgenic lineages varies between 1% and 100%, and their distribution throughout the hepatic lobule may be either central, peripheral, focal, or universal in a lineage-dependent manner, and these features breed true from generation to generation. The only meaningful variable in these experiments that can be responsible for these differences in expression is the transgene integration site. These observations imply that HBV gene expression following integration during natural HBV infection is determined not only by the structure of the integrated sequences and the differentiation state of the cell but also by the genetic and structural characteristics of the integration site itself. They also suggest that part of the transcriptional heterogeneity of hepatocytes may be due to factors other than blood flow-induced hormonal and nutrient gradients.

HBV gene expression in transgenic mice is also developmentally regulated (FARZA et al. 1987; DELOIA et al. 1989), with HBV-specific transcripts being first detected on day 14–15 of embryogenesis and increasing gradually thereafter until birth, with a dramatic increase perinatally, perhaps due to the large amounts of glucocorticoids present at that time. Indeed, injection of dexamethasone leads to a dramatic increase in serum HBsAg levels in transgenic mice (CHISARI et al. 1986; FARZA et al. 1987). This observation may rationalize the detection of a glucocorticoid-responsive element in HBV by TUR-KASPA et al. (1986) in analyses of viral expression in established cell lines.

One of the characteristics of HBV infection is its relative prevalence (ca. fivefold) among males as compared with females. Several groups (BABINET et al. 1985; FARZA et al. 1987; DELOIA et al. 1989; CHISARI et al. 1989a) have shown that HBV envelope expression is greater in male than in female transgenic mice. Indeed, liver HBV mRNA and serum HBsAg levels rise dramatically in male transgenic mice at puberty, presumably in response to sex hormonal induction. Accordingly, it has been shown (FARZA et al. 1987; DELOIA et al. 1989) that castration causes a decline in expression that is reversible by administration of androgens and, to a lesser extent, by estrogens. Thus, HBV gene expression is hormonally regulated in vivo, and it is conceivable that the male prevalence of HBV infection in the natural setting is at least partially determined at this level. The data also raise the possibility that hormonal changes such as those associated with stress, inflammation, etc. might induce changes in hepatic HBV gene expression that, in turn, could influence aspects of viral replication and pathogenesis in a heretofore unsuspected manner (see below).

4 Liver Disease in HBV Transgenic Mice

As noted at the beginning of this section, the transgenic mouse model provides us with the opportunity of studying liver disease mediated directly by the expression of viral gene products, an issue which is almost impossible to study in natural HBV infections. Moreover, because these animals are inbred it is possible to study specific components of immune-mediated injury by transplantation into transgenic mice of well-characterized immune-effector cells from nontransgenic mice primed with various viral antigens. Thus far, my laboratory has concentrated on the HBV envelope region. We have used constructs in which a HBV *Bgl* II fragment (nt 2839-1986, Fig. 1) containing the entire HBV envelope coding region, the viral enhancer, and the HBV X region has been cloned downstream of relatively liver-specific regulatory elements that are either inducible (metallothionein, MT) or highly constitutively active (albumin, Alb) in the mouse (CHISARI et al. 1985, 1986, 1987, 1989a). Within these constructs the MT or Alb promoters take the place of the pre-S promoter that normally directs synthesis of the mRNA for the large envelope protein. One lineage that contains the metallothionein-HBV envelope (MT-env) construct and 10 lineages that construct and 10 lineages that contain the albumin-HBV envelope (Alb-env) construct have been produced, and selected lineages have been characterized in detail for evidence of progressive liver damage.

4.1 Role of the HBV Large Envelope Protein as a Mediator of Liver Disease and Hepatocellular Carcinoma

When the middle and major envelope polypeptides are produced in excess, they assemble into small (22-nm), spherical particles that are formed by internal budding into the lumen of the ER and are readily secreted (EBLE et al. 1986; SIMON et al. 1988). When produced at a roughly equimolar ratio with respect to the other envelope polypeptides, the large envelope polypeptide prevents the formation of small spherical HBsAg particles and causes the formation of long, branching HBsAg filaments (containing all 3 envelope polypeptides), which become

the cellular genome, which program the cell for unrestrained growth. Thus, our analyses of the transgenic mouse lead to the hypothesis that hepatocellular carcinoma in human HBV infection may often be due to secondary genetic events entirely unrelated to the integration of HBV DNA sequences, and that the acquisition of these secondary changes is directly related to the severity, extent, and duration of the virus-induced liver cell injury, whether due to immunological mechanisms or to the accumulation of directly cytotoxic viral gene products, or to other currently undefined events. In support of this hypothesis is the fact that human hepatocellular carcinoma occurs in the context of necrosis, inflammation, regeneration, and cirrhosis in several diseases other than hepatitis, such as chronic nonA-nonB hepatitis (ALTER 1988), alcoholism (LIEBER et al. 1986), hemochromatosis (NIEDERAU et al. 1985), glycogen storage disease (LIMMER et al. 1988), alpha-1-antitrypsin deficiency (CARLSON and ERIKSSON 1985; ERIKSSON et al. 1986), and primary biliary cirrhosis (MELIA et al. 1984). It is likely that transient hepatocellular regeneration alone is not sufficient to induce these putative genetic lesions since HCC does not develop following partial hepatectomy nor does it develop in the rapidly proliferating fetal liver. Rather, prolonged injury and regeneration seem to be of major importance. Thus, additional cofactors are probably needed for the development of HCC in this setting. Such cofactors might include mutagenic products of the associated inflammatory cells or impaired detoxification or DNA repair mechanisms within the injured hepatocyte as illustrated in Fig. 2.

4.2 Immunopathogenesis of Hepatocellular Injury

The HBV transgenic mouse model has created the opportunity to study not only liver cell injury mediated by expression of viral gene products but also to characterize the pathogenetic potential of the HBV-specific cytotoxic T cell (CTL) response. The role of CTLs in human hepatitis B has been inferred from the isolation from the liver (biopsies) and blood of helper and suppressor T cells reactive to specific viral antigens. It has not been possible, however, to demonstrate that these human lymphocytes actually play a role in liver injury. These issues can be studies in transgenic mice into which specific effector cells can be introduced by adoptive transfer (see below).

The minimal experimental model for T-cell-mediated liver cell injury requires hepatocellular synthesis of HBV gene products, the appropriate processing and association of these products with major histocompatibility determinants within the hepatocyte, the expression of the MHC-HBV peptide complex on the hepatocyte surface, the induction of a cellular immune response specific for one or more of the MHC-associated HBV peptides (i.e., the T-cell repertoire must include rearranged variable region genes specific for the expressed antigens, and it must be effectively induced), the appropriate functional T-cell subset must be expanded, and the target cell, i.e., the primary hepatocyte, must be susceptible to T-cell-mediated cytolysis. In view of the complexity of these requirements it is not surprising that the immunopathogenesis of HBV-induced liver disease is not yet defined.

Using the transgenic mouse model, we have demonstrated that HBV antigen-specific immune effector mechanisms can destroy HBV antigen-positive hepatocytes in vitro and, more importantly, in vivo (MORIYAMA et al. 1990). In these experiments, we used a transgenic mouse lineage (107-5) that expresses noncytotoxic amounts of the HBV large and major envelope polypeptides in the hepatocytes and, hence, does not develop spontaneous liver disease. (Lineage 107-5, like those that do overexpress L protein, contains a transgene in which the albumin promoter was placed in front of the HBV envelope.) This lineage develops liver cell injury, however, following the adoptive transfer of HBsAg-primed spleen cells or HBs-specific antibody from nontransgenic, syngeneic mice that have been immunized by infection with vaccinia-HBV envelope recombinants (MORIYAMA et al. 1990).

Injury mediated by HBs-specific antibody is rapid in onset and coincides with the disappearance of HBsAg and the appearance of anti-HBs in the serum. Clearly, this could be due either to complement-mediated cell lysis or to antibody-mediated cellular cytotoxicity (ADCC). A delayed form of liver cell injury occurs several days after the adoptive transfer of HBs antigen-primed spleen cells and is characterized histologically by widespread focal hepatocellular necrosis and a lymphomononuclear cell infiltrate with associated lymphoid infiltration in portal areas, thus resembling the histological picture of chronic hepatitis. In studies with fractionated spleen cell populations, we found that liver cell injury was mediated only by the T-cell population, specifically by the Lyt2 $(CD8)^+$ T-cell subset. A $CD8^+$, MHC class I-restricted, cytotoxic T-cell line that displayed HBV envelope-specific cytotoxicity was then established. The fine specificity of the CTL response was mapped to an epitope located between amino acids 21 and 40 of the major envelope polypeptide (Fig. 1). The HBsAg-specific CTL line caused liver cell injury when transferred into the transgenic mice in vivo. Injury was detectable as early as 5 days post transfer, and it persisted for several weeks thereafter, decreasing as a function of time. The data indicate that HBsAg-specific, MHC class I-restricted, $CD8^+$ CTL are functional in vivo and that endogenously synthesized HBV envelope antigens are expressed at the cell surface in a form that is recognized by such CTL.

Thus, studies with transgenic mice establish the scientific basis for the hypothesis that HBV-induced liver disease is mediated both as a direct effect of cytotoxicity associated with viral protein synthesis, especially aberrant overexpression of the large envelope protein, and as a consequence of immune reactivity to viral envelope antigens expressed on infected hepatocytes. Obviously, the immunogenicity of the remaining HBV-encoded antigens with respect to the induction of CTL and the in vivo susceptibility of the corresponding transgenic hepatocytes to destruction by such CTL against a variety of MHC backgrounds should be assessed to understand fully the potential contribution of the response to each HBV antigen with respect to viral clearance and liver cell injury in viral hepatitis.

5 HBV Gene Expression During the Acute Phase Response

Much of what is known about expression of the HBV genome is derived from studies in which the viral genes and regulatory elements have been introduced into mammalian cells which may bear little resemblance to the primary hepatocyte in which the virus normally exists. Furthermore, although viral hepatitis is an inflammatory process, nothing is known about the influence of inflammatory cells and their products on HBV expression by the liver cell. This is important because inflammation characteristically induces profound alterations in the expression of several hepatocellular proteins (acute phase response) including C-reactive protein, fibrinogen, serum anyloid A, alpha-2-macroglobulin, haptoglobin, albumin, metallothionein, alpha-1-antitrypsin, C3, etc. The acute phase response is mediated by inflammatory cell cytokines, most notably tumor necrosis factor, interleukin-1, interleukin-6, and gamma-interferon (FEY and GAULDIE 1989), which regulate hepatocellular gene expression at the transcriptional and posttranscriptional levels (FEY and GAULDIE 1989).

Since the HBV genome is exposed to the same inflammatory mediators as endogenous hepatocellular genes during acute and chronic hepatitis, it is possible that HBV gene expression might also be modulated during the acute phase response, and that such changes could influence the evolution of the disease. Clearly, this is an important aspect of the virus-host interaction that deserves attention.

Since the acute phase response can be induced experimentally by the administration of bacterial lipopolysaccharide (LPS), we (P. GILLES and F.V. CHISARI, unpublished observations) analyzed HBV gene expression in transgenic mice following the intraperitoneal administration of LPS under conditions known to induce an acute phase response (GLIBETIC and BAUMANN 1986). HBV-specific gene expression (steady state mRNA) was substantially decreased after LPS administration concomitant with an increase in transcripts corresponding to the acute phase proteins SAA and alpha-2-macroglobulin. The molecular basis for these effects is currently under investigation.

Because of the implications of these observations with respect to hepatocellular viral gene expression in the inflamed liver and the derivative implications with respect to the pathogenesis of acute hepatitis, chronic hepatitis, and the healthy carrier state, this area warrants further investigation.

6 Conclusions

Despite recent advances, we are left with major areas of ignorance with respect to the biology and pathogenesis of HBV that the transgenic mouse model can be used to resolve. Its great strength rests in the ability to express each of the viral genes individually and to explore the tissue specificity of every viral regulatory element in all of the organs of the living animal. Indeed, the individual functions of the nucleocapsid, polymerase, and X ORFs remain to be examined in this model. The consequences of an immune response to each of the viral gene products can be studied in this system such that the relative importance of each of the viral antigens and the individual components of the immune response can be assessed in terms of viral clearance, liver cell injury, and hepatocellular carcinoma. The mechanisms responsible for immunological tolerance can be probed in an effort to understand the basis for the establishment of the chronic carrier state in humans. An important objective is the introduction of the HBV receptor into transgenic mice under the control of liver-specific control sequences to permit the creation of the ultimate animal model of HBV infection and the examination of issues relating to vaccine development, antiviral therapy, viral immunobiology, and pathogenesis that are otherwise unapproachable at this time.

Acknowledgments. I thank Takashi Moriyama Patrick Gilles, John Shutter, Pierre Filippi, Carl Pinkert, Richard Palmiter, and Ralph Brinster for collaboration in the production and analysis of transgenic mice. I thank Darlene Corrigan for manuscript preparation. This research was supported by NIH grants CA40489, CA38635, AI20001, AI20720, HD17321, HD0912, and HD07155 and by a North Atlantic Treaty Organization Grant for International Collaboration (675/84). This is publication number 6246-MEM from the Department of Molecular and Experimental Medicine, Research Institute of Scripps Clinic, La Jolla, CA, USA.

References

Alter HJ (1988) Transfusion-associated non-A, non-B hepatitis. The first decade. In: Zuckerman AJ (ed) Viral hepatitis and liver disease. Liss, New York, pp 537–542

Araki K, Miyazaki J, Tsurimoto T, Inomoto T, Iwanaga T, Matsubara K, Yamamura K (1989a) Demethylation by 5-azacytidine results in the expression of hepatitis B virus surface antigen in transgenic mice. Jpn J Cancer 80: 295–298

Araki K, Miyazaki J, Hino O, Tomita N, Chisaka O, Matsubara K, Yamamura K (1989b) Expression and replication of hepatitis B virus genome in transgenic mice. Proc Natl Acad Sci USA 86: 207–211

Babinet C, Farza H, Morello D, Hadchouel M, Pourcel C (1985) Specific expression of hepatitis B surface antigen (HBsAg) in transgenic mice. Science 230: 1160–1163

Bulla G, Siddiqui A (1988) Negative regulation of the hepatitis B virus pre-S1 promoter by internal DNA sequences. In: Abstracts of papers presented at the 1988 meeting on molecular biology of hepatitis B Virus, La Jolla. University of California, San Diego Aug 7–11, 1988; p 64

Burk RD, DeLoia JA, ElAwady MK, Gearhart JD (1988) Tissue preferential expression of the hepatitis B virus (HBV) surface antigen gene in two lines of HBV transgenic mice. J Virol 62: 649–654

Buscher M, Reiser W, Will H, Schaller H (1985) Transcripts and the putative RNA pregenome of duck hepatitis B virus implications for reverse transcription. Cell 40: 717–724

Carlson J, Eriksson S (1985) Chronic "cryptogenic" liver disease and malignant hepatoma in intermediate alpha-1-antitrypsin deficiency identified by a PiZ-specific monoclonal antibody. Scand J Gastroenterol 20: 835–842

Chisari FV, Pinkert CA, Milich DR, Filippi P, McLachlan A, Palmiter RD, Brinster RL (1985) A transgenic mouse model of the chronic hepatitis B surface antigen carrier state. Science 230: 1157–1160

Chisari FV, Filippi P, McLachlan A, Milich DR, Riggs M, Lee S, Palmiter RD, Pinkert CA, Brinster RL (1986) Expression of hepatitis B virus large envelope polypeptide inhibits hepatitis B surface antigen secretion in transgenic mice. J Virol 60: 880–887

Marquardt O, Heermann KH, Seifer M, Gerlich WH (1987) Cell type specific expression of pre S1 antigen and secretion of hepatitis B virus surface antigen. Arch Virol 96: 249–256
McLachlan A, Milich DR, Raney AK, Riggs MG, Hughes JL, Sorge J, Chisari FV (1987) Expression of hepatitis B virus surface and core antigens: influences of pre-S and precore sequences. J Virol 61: 683–692
[illegible] WM, Johnson RG, [illegible] (1984) Hepatocellular carcinoma in [illegible]. Gastroenterology 87: 660–667
[illegible] (1989) HBcAg expressed on the surface of circulating [illegible] particles in patients with [illegible] hepatitis B virus infection without evidence of anti-HBc formation. Hepatology 10: 179–185
Molnar-Kimber KL, Jarocki-Witek V, Dheer SK, Vernon SK, Conley AJ, Davis AR, Hung PP (1988) Distinctive properties of the hepatitis B virus envelope proteins. J Virol 62: 407–416
Moriyama T, Guilhot S, Klopchin K, Moss B, Pinkert CA, Palmiter RD, Brinster RL, Kanagawa O, Chisari FV (1990) Immunobiology and pathogenesis of hepatocellular injury in hepatitis B virus transgenic mice. Science 248: 361–364
[illegible] (1985) [illegible]
Ou J, Rutter WJ (1985) Hybrid hepatitis B virus-host transcripts in a human hepatoma cell. Proc Natl Acad Sci USA 82: 83–87
Ou JH, Rutter WJ (1987) Regulation of secretion of the hepatitis B virus major surface antigen by the preS-1 protein. J Virol 61: 782–786
Ou JH, Yeh CT, Yen TSB (1989) Transport of hepatitis B virus precore protein into the nucleus after cleavage of its signal peptide. J Virol 63: 5238–5243
Patzer EJ, Nakamura GR, Yaffe A (1984) Intracellular transport and secretion of hepatitis B surface antigen in mammalian cells. J Virol 51: 346–353
Patzer EJ, Nakamura GR, Simonsen CC, Levinson AD, Brands R (1986) Intracellular assembly and packaging of hepatitis B surface antigen particles occur in the endoplasmic reticulum. J Virol 58: 884–892
Pourcel C (1989) Souris transgéniques: nouveaux modèles d'études du virus de l'hépatite B. Médecine Sciences 5: 245–249
Rall LB, Standring DN, Laub O, Rutter WJ (1983) Transcription of hepatitis B virus by RNA polymerase II. Mol Cell Biol 3: 1766–1773
Schlicht HJ, Schaller H (1989) The secretory core protein of human hepatitis B virus is expressed on the cell surface. J Virol 63: 5399–5404
Seto E, Yen TSB, Peterlin BM, Ou JH (1988) Trans-activation of the human immunodeficiency virus long terminal repeat sequence by the hepatitis B virus X protein. Proc Natl Acad Sci USA 85: 8286–8290
Siddiqui A, Jameel S, Mapoles J (1986) Transcriptional control elements of hepatitis B surface antigen gene. Proc Natl Acad Sci USA 83: 566–570
Simon K, Lingappa VR, Ganem D (1988) Secreted hepatitis B surface antigen polypeptides are derived from a transmembrane precursor. J Cell Biol 107: 2163–2168
Spandau DF, Lee CH (1988) Trans-activation of viral enhancers by the hepatitis B virus X protein. J Virol 62: 427–434
Standring DN, Ou JH, Rutter WJ (1986) Assembly of viral particles in Xenopus oocytes: pre-surface-antigens regulate secretion of the hepatitis B viral surface envelope particle. Proc Natl Acad Sci USA 83: 9338–9342
Tur-Kaspa R, Burk RD, Shaul Y, Shafritz DA (1986) Hepatitis B virus DNA contains a glucocorticoid-responsive element. Proc Natl Acad Sci USA 83: 1627–1631
Twu JS, Schloemer RH (1987) Transcriptional trans-activating function of hepatitis B virus. J Virol 61: 3448–3453
Twu JS, Chu K, Robinson WS (1989) Hepatitis B virus X gene activates κB-like enhancer sequences in the long terminal repeat of human immunodeficiency virus 1. Proc Natl Acad Sci USA 86: 5168–5172
Will H, Reiser W, Weimer T, Pfaff E, Büscher M, Sprengel R, Cattaneo R, Schaller H (1987) Replication strategy of human hepatitis B virus. J Virol 61: 904–911
Wollersheim M, Debelka U, Hofschneider PH (1988) A transactivating function encoded in the hepatitis B virus X gene is conserved in the integrated state. Oncogene 3: 545–552
Yamamura K, Tsurimoto T, Ebihara T, Yamamoto K, Fujiyama A, Chisaka O, Matsubara K (1987) Methylation of hepatitis B virus DNA and liver specific suppression of RNA production in transgenic mice. Jpn J Cancer Res 78: 681–688

Cellular and Molecular Mechanisms of Hepatocarcinogenesis Associated with Hepadnavirus Infection

C. E. Rogler

Liver Research Center, Department of Medicine, Albert Einstein College of Medicine, 1300 Morris Park Avenue, Bronx, NY 10401, USA

Current Topics in Microbiology and Immunology, Vol. 168

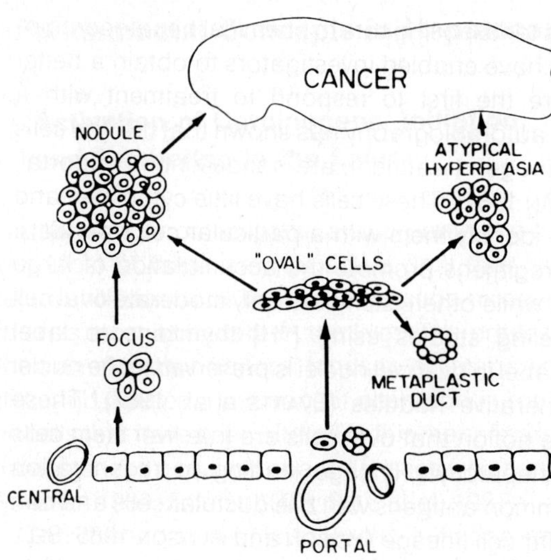

Fig. 1. Possible cellular lineages in experimental hepatocarcinogenesis in rats (adapted from SELL et al. 1987). Possible relationships between putative premalignant cellular changes in the livers of rats exposed to chemical carcinogens and the malignant tumor that eventually appears are depicted. Liver stem cells present in the portal region proliferate and appear as "oval" cells which differentiate to form atypical hyperplasia and/or nodules. These progress to form hepatocellular carcinoma. Carcinoma may also arise from normal hepatocytes that become altered and proliferate to form a small "focus" that may progress to nodules and then to cancer

carcinogen treatment transform the liver into a highly susceptible target for carcinogenesis. In the remainder of this review we will discuss how hepadnaviruses provide both the mitotic stimulus and the carcinogenic agent which are necessary for the development of HCC.

2 Hepadnavirus Infection and HCC

2.1 Acute HBV Infection in Relation to the Carcinogenic Mechanisms

During acute HBV infection, both degenerative and regenerative changes coexist. In acutely infected individuals the development of a complete immune response is sufficient to eliminate viral replication completely, allowing recovery to proceed. During the recovery phase, liver regeneration is prominent and is manifested by anisonucleosis, bi- and trinucleate cells, and mitoses (HOLLINGER 1985a).

In terms of carcinogenic models, the liver injury and regeneration which occur during acute HBV infection can serve to generate initiated hepatocytes, since acutely infected individuals have an elevated risk of developing HCC. The number of initiated hepatocytes produced probably depends on the severity of the hepatitis and the length of time the infection persists. The presence of initiated hepatocytes also represents an irreversible step towards susceptibility to HCC. However, the initiated hepatocytes will not progress toward malignancy without the further action of tumor promoters. Thus, individuals experiencing only acute infections are still at much lower risk than those which develop persistent infection (BEASLEY et al. 1981).

2.2 Persistent HBV Infection and Hepatocarcinogenesis

It was previously thought that acute HBV infection progresses to persistent infection in 5%–10% of adult HBV patients (HOLLINGER 1985a; BEASLEY et al. 1981). The use of Polymerase Chain Reaction (PCR) methodology has shown that HBV infection may persist in a much greater percentage of individuals. The course of chronic liver disease caused by persistent HBV infection is quite varied. In the case of asymptomatic HBV carriers, liver disease may be barely detectable for years. Alternatively, persistent infection may cause a relentless progression to cirrhosis and death (HOLLINGER 1985b). In regard to carcinogenic risk, even these persistently infected individuals, who may be completely asymptomatic, are probably exposed to both a continuous carcinogenic agent (the hepadnavirus) and a mitogenic stimulus (generated by a limited immune response) placing them at very high risk of developing HCC. In fact, male HBV carriers have a 40% lifetime risk of HCC (BEASLEY et al. 1981). The results do not, however, rule out the possibility that some carriers are actually at very low risk because of the very low level of mitogenic stimulus.

In humans, several stages in the progression from mild persistent hepatitis to chronic active hepatitis (CAH) and finally to HCC have been observed. In most HBV carriers, replication gradually ceases over the course of many years while hepatocytes continue to express viral envelope proteins. RAIMONDO and colleagues (RAIMONDO et al. 1988) have also found that viral DNA replication may continue in some patients in the absence of virus production. These findings are all consistent with a breakdown in the ability of hepatocytes to support a critical event in virus maturation or gene expression subsequent to viral DNA synthesis (MASON and TAYLOR 1989) and suggest that infection-associated injury induces dramatic changes in the hepatocyte population, perhaps including a selection for hepatocytes that are no longer able to support virus replication.

In fact, in many individuals active virus replication eventually either ceases or declines to very low levels, with concurrent loss of HBeAg from the serum and the appearance of HBe-specific immunoglobulins (chapter by D. GANEM, this volume). The observation of increased liver disease activity during this transition period suggests a major loss of infected hepatocytes, and their replacement via regeneration with hepatocytes that may be nonpermissive for virus replication. This regenerative activity could, of course, greatly expand the pool of cancer precursor cells, placing the individual at much greater risk for HCC than that which existed immediately prior to seroconversion.

2.3 Animal Models of Hepadnavirus Infection

Four hepadnaviruses have been identified in addition to HBV, including woodchuck hepatitis virus (WHV) (SUMMERS et al. 1978), ground squirrel hepatitis virus (GSHV) (MARION et al. 1980), duck hepatitis B virus (DHBV) (MASON et al. 1980), and heron hepatitis B virus (HHBV) (SPRENGEL et al. 1988). Of these animal viruses, WHV infections are being most actively studied to gain insights into HBV infections in humans. Therefore, WHV infections and their associated pathogenic consequences will be discussed to the greatest extent in the following sections.

2.3.1 Woodchuck Hepatitis Virus

Acute WHV infection in the liver of woodchucks is characterized by focal eosinophilic changes of the cytoplasm of hepatocytes. Activation of sinusoidal cells occurs, and portal tracts show extensive infiltration by lymphocytes. Lymphocytes and plasma cells also infiltrate the liver parenchyma but to a lesser extent than in the portal tracts (POPPER et al. 1987). At the height of acute lesions, proliferated bile ductules are present in the periportal areas (POPPER et al. 1987; KORBA et al. 1987). Within the context of carcinogenic models, these acute lesions may provide the mitogenic stimulus needed to propagate mutation in initiated cells that may be present in the liver as a result, perhaps, of events unrelated to virus infections (but see below). The existence of proliferating bile ductules and newly regenerated hepatocytes in acutely infected woodchuck liver (POPPER et al. 1987) is consistent with the possibility that liver stem cell proliferation is responsible for part, if not all, of the restoration of liver mass in animals recovering from acute infection.

Several cases of HCC have indeed been reported in woodchucks which experienced acute WHV infection followed by seroconversion to WHc- and WHs-specific antibodies (KORBA et al. 1987). The specific mechanism by which WHV served as a carcinogenic agent in these cases is unknown. However, the tumors in these animals contained WHV DNA integrations, and since integration is a rare event during acute infection (see later section) these results suggest that cells containing integrated WHV may have been selectively targeted, or placed at higher risk, for future malignant transformation possibly involving N-*myc* activation (see Sect. 5.2).

Persistent WHV infection is commonly characterized by a state of mild lymphocytic infiltration in the portal tracts known as portal hepatitis (PH) (POPPER et al. 1987; KORBA et al. 1987). In PH, hepatocytes which are analogous to "ground glass" cells in HBV carriers are frequent (see chapter by F.V. CHISARI, this volume, for review). This minimal state in woodchucks corresponds to the human "healthy" HBsAg carrier state. Mild PH develops most frequently in neonatal rather than adult infections of woodchucks and presents a high risk of HCC (POPPER et al. 1987; SNYDER and SUMMERS 1980); adult woodchucks usually develop acute infections which are resolved. In the most carefully controlled study conducted so far, 8 of 8 (100%) captive woodchucks infected at birth with WHV and maintained in captivity developed HCC after 17–36 months of

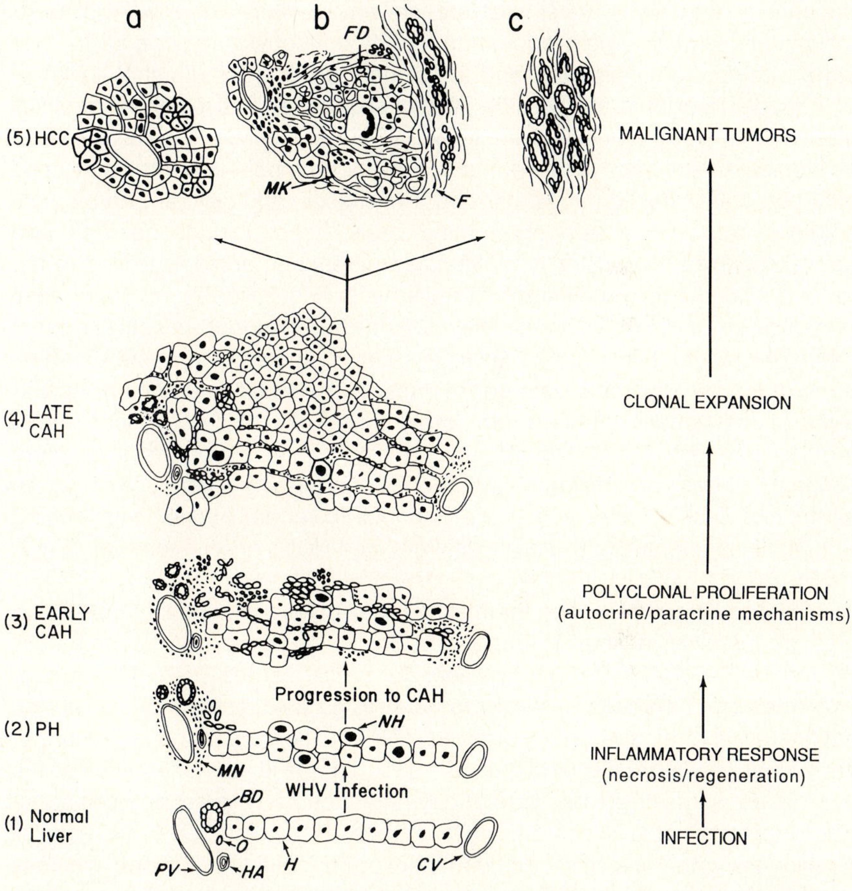

Fig. 2a–c. The proposed progressive stages of WHV infection leading to the generation of biochemically and histologically heterogeneous primary hepatocellular carcinoma (HCC) in woodchucks. *O*, oval cell; *H*, hepatocyte; *BD*, bile duct; *CV*, central vein; *PV*, portal vein; *HA*, hepatic artery; *MN*, mononuclear cell; *F*, fibroblast; *MK*, megakaryocyte; *FD*, fatty degeneration; *NH*, necrotic hepatocyte. *1*, Schematic illustration of normal liver architecture comprising portal tracks (*left*) connected by single cell plates of hepatocytes and central hepatic veins (*right*). *2*, After establishment of WHV persistent infection, portal hepatitis (*PH*) develops. *3*, Portal hepatitis progresses to chronic active hepatitis (*CAH*). Oval cell proliferation is prominent. *4*, Precancerous nodules develop throughout the liver parenchyma as oval cell proliferation and an inflammatory response continue. *5*, Sustained by the dynamics of protracted infection, inflammation, cell necrosis, and growth factor production, heterogeneous types of tumors arise. Illustrated are; (**a**) a highly differentiated "adult type" HCC, (**b**) a poorly differentiated, anaplastic, "fetal type" HCC with extensive extramedullary hematopoesis, and (**c**) a rare cholangiocarcinoma

persistent infection. This represents the highest reported incidence of HCC obtained following insult with any carcinogenic agent (POPPER et al. 1987).

During mild PH, oval cells are present in the portal tracts of WHV-infected woodchucks in low but elevated numbers compared with their very rare occurrence in normal liver (FU et al. 1988). However, a period of necroinflammation is almost always observed in the precancerous stages preceding and concurrent with the development of HCC (POPPER et al. 1987). During this precancerous period oval cell proliferation may become extensive, with focal accumulations of oval cells appearing in the centrilobular regions of the liver along with cords of cells exhibiting transitional phenotypes (FU et al. 1988).

A model to explain the possible relationship of reparative cellular proliferation and the eventual occurrence of HCCs with specific biochemical and histological phenotypes is presented in Fig. 2. The model proposes that once WHV persistent infection is established, PH ensues, and several pathological reactions occur on a self-perpetuating basis. These include necrosis and regeneration of hepatocytes, an inflammatory reaction primarily limited to the portal tracts, and mild proliferation of bile ducts (Fig. 2, line 2). As the disease progresses to CAH, involvement of the rest of the hepatic lobule becomes evident (lobular hepatitis) (Fig. 2, line 3). Oval cells, either isolated or in clusters, can be found along the sinusoids from portal tracts to the midzone and centrilobular areas of the hepatic lobule. From within this milieu of cells, precancerous nodules develop (Fig. 2, line 4), and selected nodules progress to more malignant phenotypes through the autocrine and paracrine activities of growth factors (see Sect. 5). In the final stage, tumors with various phenotypes, ranging from highly differentiated to poorly differentiated HCCs arise in the liver (Fig. 2, line 5).

Therefore, one explanation for the heterogeneity of tumor phenotypes which arise in the liver is that transitional cells which have not become fully differentiated hepatocytes are present in precancerous liver. The further differentiation of these cells is likely to be arrested as a result of, or as part of, malignant transformation since arrested differentiation is a common feature of many types of tumor cells (HEARD et al. 1984). Thus, the final phenotype of the tumor most likely reflects the state of differentiation of the precursor cell at the time of transformation. According to this model, transformation of transitional hepatocytes generates less differentiated, more fetal-like tumors, whereas transformation of cells further along the hepatocyte lineage produces a more highly differentiated tumor.

2.3.2 Ground Squirrel Hepatitis Virus and Duck Hepatitis B Virus

Experimental infection of ground squirrels with GSHV results in either (1) self-limited or transient acute infection, (2) a persistent infection, or (3) antibody production without detectable viral replication (MARION et al. 1983). The predominant histologic reaction during persistent infection is a mild hepatitis. Some portal tracts are expanded by infiltration with mononuclear inflammatory cells which surround the bile ducts and ductules. Proliferated bile ductules are present, and inflammation sometimes extends into the surrounding parenchyma, replacing hepatocytes in a few places (MARION et al. 1983).

HCC takes longer to develop in ground squirrels than woodchucks, perhaps reflecting the fact that a mild hepatitis is retained for longer periods, rather than progressing as in the woodchuck (MARION et al. 1986). Nonetheless, in older ground squirrels, with persistent GSHV infection, there is a very high rate of progression to HCC, similar to that in woodchucks (MARION et al. 1986). HCC may also develop in ducks as a result of persistent infection with DHBV (YOKOSUKA et al. 1985). Histologic analysis of persistent infection in ducks reveals an even milder hepatitis than in ground squirrels (OMATA et al. 1983). Consistent with this, HCC may take even longer to develop in persistently infected ducks than in ground squirrels (YOKOSUKA et al. 1985).

3 Mechanisms of Hepatocellular Injury —Maintenance of the Mitogenic Stimulus

Since cell death and regeneration are essential parts of the carcinogenic process, it is important to understand the mechanisms by which viral infection promotes these activities. Since most infected hepatocytes appear healthy during persistent infection, and viral replication can be maintained in tissue culture without cell death, it is evident that hepadnaviruses are not normally cytotoxic to their host cells. Hepatocellular injury is therefore mediated by the host immune system and under certain circumstances by abnormal production of various viral proteins.

3.1 Immunopathogenesis of Persistent HBV Infection

The primary issues regarding immunopathogenesis during persistent HBV infection involve the role of the cytotoxic T-lymphocyte (CTL) response and of the B-cell-mediated immune response as mediators of liver damage. Studies using mouse models have provided significant insights into the role of T-cell-independent and T-cell-dependent immune responses in the mouse (MILICH and CHISARI 1982; MILICH et al. 1985, 1986, 1987), and these studies provide a basis for further investigations of the role of these responses in woodchucks and humans. The degree to which various viral proteins contribute to the CTL response or to antibody-dependent cellular cytotoxicity is still a matter of active investigation. The CTL response could be directed to several HBV-encoded antigens displayed on the hepatocyte membrane including the envelope, nucleocapsid, polymerase, or even the X polypeptides. The failure to generate antibodies to HBs antigens is believed to be one of the factors required for the maintenance of

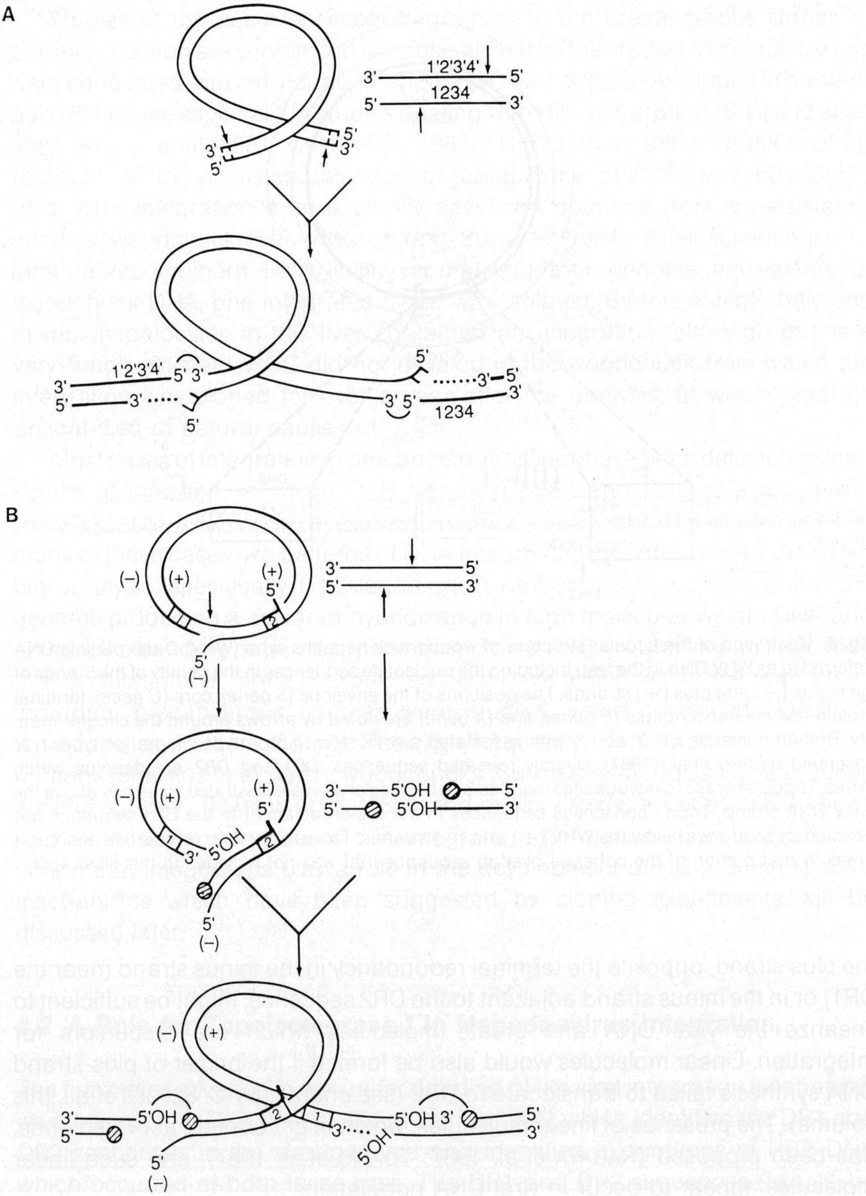

Fig. 4 A, B. Illustration comparing retrovirus integration with the proposed topoisomerase I-mediated hepadnavirus integration mechanism. **A** Retroviral integration. A linear, double-stranded, viral DNA molecule is nicked 2 bp from its 3′ end producing two unpaired bases at each 5′ terminus (double-stranded loop). This molecule is integrated at a cellular DNA site, which is nicked at staggered positions through the action of the viral integrase followed by joining of the 5′ ends of cellular DNA to the 3′ ends of viral DNA. The 2-base 5′ overhang is lost from each end of the viral DNA during integration. The complete integration reaction therefore generates a short direct duplication of cellular DNA at each viral-cell junction. **B** Hepadnavirus integration. Topo I linearizes open circular

covalently bound to DNA via a 3′ phosphotyrosyl bond while the DNA swivels (CHAMPOUX 1977, 1978). If a proper 5′OH acceptor end is provided, these 3′ bound Topo I molecules are capable of linking heterologous DNA fragments via a nonhomologous recombination mechanism (HALLIGAN et al. 1982).

A survey of HBV and WHV sequences across integration sites has shown that preferred Topo I cleavage sites are present at or near 91% of the HBV integration sites cloned from HCC (SCHIRMACHER and ROGLER, unpublished data and Table 1). This value is virtually identical with a previous survey of DNA sequences across sites of illegitimate recombination in which Topo I sites were present in 92% of the cases (KONOPKA 1988). In summary, these data support the notion that Topo I plays a role in HBV integration in vivo.

Recent studies have demonstrated that Topo I cleaves WHV DNA at a cluster of bases near the DR1 sequence in plus-strand DNA and that these sites are located within the cohesive overlap of virion DNA (Fig. 3) (WANG and ROGLER 1991). In addition to cleavage, Topo I can link viral DNA to the ends of heterologous cellular DNA fragments in in vitro reactions (WANG and ROGLER 1991). A new "integration" function to Topo I, in which Topo I mediated integration of WHV virion DNA into an internal site in double-stranded cellular DNA, was also described in this study. The in vitro generated integrations closely resemble in vivo integration in that they show the same end specificity and polarity as in vivo integrations (WANG and ROGLER 1991).

Taking into consideration the Topo I cleavage mechanism, a prototype model for the integration of both ends of a Topo I linearized hepadnaviral molecule has been proposed (WANG and ROGLER et al. 1991), and this model is compared with the integration model of retroviral DNA (GRANDGENETT and MUMM 1990) in Fig. 4. In retroviruses, the integration precursor is a linear DNA molecule. Each strand of the linear molecule is specifically nicked 2 bp from its 3′ end, and the immediate precursor molecule therefore contains two unpaired bases at its 5′ termini. This molecule becomes integrated at a cellular site which is also nicked at positions which produce cohesive ends with 5′ overhangs. The complete integration reaction generates a short direct duplication of cellular DNA due to the separation of the cellular cohesive ends at the integration site (Fig. 4).

The hepadnavirus integration model resembles the retroviral mechanism in some respects but is unique in many others. It is similar in that a linearized viral molecule is the immediate precursor for integration, and cellular DNA is cut at

virion DNA by nicking the plus strand opposite the terminal redundancy in the minus strand (at the fork in virion DNA). It also cleaves cellular DNA at staggered positions. It remains covalently bound to the 3′ phosphate of the viral DNA plus strand and mediates integration, by linking it to a free 5′ OH in the cellular DNA. The opposite end of the plus strand contains a free 5′ OH group which serves as an acceptor site for a 3′ Topo I molecule bound to cellular DNA. The directly repeated (DR1, DR2) sequences in hepadnavirus DNA are noted by *enclosed boxes* with *1* and *2*, respectively. *Circles with hash marks* are Topo I molecules. *Arrows* immediately adjacent to DNA molecules are nicking sites. *Thick squiggly line* opposite 2 is the plus-strand RNA primer. *Dots*, regions of the molecules which must be filled in by cellular DNA polymerases staggered positions in opposite strands. However, in the case of hepadnaviruses, integration is usually associated with microdeletions of cellular DNA instead of direct duplications (NAKAMURA et al. 1988; NAGAYA et al. 1987; HINO et al. 1989). One case of a direct duplication has been reported.

According to the hepadnavirus integration model, Topo I is responsible for linearizing circular virion DNA, and it remains bound to the 3′ end of the viral molecule after cleavage and linearization (Fig. 4). These Topo I-containing ends function as "donors" during integration and rearrangement reactions. The opposite end of the viral molecule contains a free 5′OH which can function as an acceptor site for other DNA ends with 3′ bound Topo I molecules. In order for a linearized viral molecule to integrate, it must insert into the reaction site at the moment of Topo I nicking and provide an alternate acceptor or donor end. The complexity of such reactions may account for the low frequency of integration events observed in vivo. On the other hand, reactions which involve intramolecular joining of Topo I-cleaved ends of the same molecule would be expected to occur more often and would generate rearranged molecules. Most integrations cloned from HCC have inversions and deletions, possibly due to Topo I-mediated rearrangements, some of which may have occurred prior to integration into cellular DNA (ROGLER and SUMMERS 1982, see Sect. 4.4).

The finding that Topo I mediates the integration of open circular DNA (WANG and ROGLER 1991) may have significance with regard to the maintenance of persistent infection. For example, if Topo I functions to divert open circular viral molecules from their normal processing to CCC DNA, the amount of viral DNA available as template for transcription could be lowered. Such a Topo I-mediated rearrangement-integration pathway would also lead to the production of defective, rearranged viral DNAs which would be incapable of replication. Molecules fitting these criteria, termed "novel forms," have been cloned from chronically infected woodchuck livers (ROGLER and SUMMERS 1982). Such a diversion pathway could also explain the gradual reduction of virus replication during persistent infection (RAIMONDO et al. 1988) and the build-up of viral integrants in the liver of long-term HBV carriers (SHAFRITZ et al. 1981). Viral CCC DNA itself may, of course, be another substrate for integration via the pathway suggested in Fig. 4.

4.3 Insertional Mutagenesis of Cellular Genes Involved in Growth Control

One of the primary goals of experiments designed to clone integrations over the past decade has been the identification of new genes involved in hepatic oncogenesis. The initial approach utilized cellular flanking sequences, adjacent to integrations from single integration tumors, as probes to search for common integration sites in tumors. The rationale for these experiments was that insertion of a viral transcriptional element adjacent to or near a cellular protooncogene would cause activation of the gene in a large subset (if not all) of HCC and therefor allow identification of protooncogenes involved in hepatocarcinogenesis. This approach was reasonable since the work of HAYWARD et al. (1981) had demonstrated that viral long terminal repeats (LTR) of avian leukosis virus activated the c-*myc* gene by a promoter insertion mechanism in bursal lymphomas of chickens.

Several significant cases of integration into cellular genes involved in growth control have also been discovered, and these may prove to be important in understanding the progression to HCC. In two cases, HBV integrations have identified new human genes with potential regulatory properties (DEJEAN et al. 1986, WANG et al. 1990), and in several other cases WHV integration occurred within and adjacent to the woodchuck c-*myc* and N-*myc* genes (MOROY et al. 1986; HSU et al. 1988; FOUREL et al. 1990). Whether the integrations in these genes were sufficient to initiate tumorigenesis or functioned to promote tumor progression awaits further experimentation. A brief description of these interesting cases is presented here.

One case of HBV integration into an intron of a human cyclin A gene has been reported (WANG et al. 1990). The integrated HBV DNA contained an HBV enhancer element, and it was speculated that this element could have been responsible for the increased steady state level of cyclin A mRNA observed in the tumor. Cyclins are characterized by their steady accumulation throughout interphase until the G2/m transition, followed by rapid disappearance at the onset of anaphase. Cyclins complex with and activate the serine-threonine kinase p34cdc2 during the G2/m phase transition (DRAETTA et al. 1989; DRAETTA and BEACH 1988) and are thought to be necessary for the initiation of mitosis and meiosis (MINSHULL et al. 1989). Their possible role in tumorigenesis could be through the control of phosphorylation of regulating proteins such as p60c-src or the retinoblastoma gene product, which is a potential inhibitor of cell proliferation (LUDLOW et al. 1989). In the specific case reported, the inserted HBV DNA contained a poly A addition site and a strong liver-specific promoter in addition to the enhancer element. These signals could have caused the production of amino-terminal and carboxyl-terminal truncated cyclin proteins of unknown (possibly deregulated) function.

In another instance HBV integration interrupted the coding region of the retinoic acid receptor B gene (DEJEAN et al. 1986; DETHE et al. 1987). In this case, integration occurred in an exon of the gene and created a chimeric viral-host open reading frame (ORF). The viral preS 1 promoter was present in the HBV sequences and served as the promoter for a viral-cellular cotranscript. Retinoic acid receptor proteins are nuclear DNA binding proteins which mediate the effects of retinoic acid (MCBURNEY et al. 1988). In some systems, retinoic acid is involved in cellular differentiation; however, its function in the precancerous liver or in HCC is unknown. Several cases are known in which mutant receptor genomes have either potentiated transformation or acted as transforming genes themselves.

The strongest evidence directly linking viral DNA integration to hepatocellular transformation is the discovery of frequent integration of WHV

Table 1. Microdeletrons that have occurred in association with HBV integrations and preferred Topo I cleavage sites (underlined)

Integration Number	Clone	Topo I Sites	5′	R	Deletion	R	3′	Topo I Sites	Reference
1	λDT1	4	CCATC GGTAG	TC AG	cacttcc gtgaagg	TCC AGG	CCCTC GGGAG	3	DEJEAN et al. (1986)
2	HC1701-1	4	CAAAA GTTTT	T A	gagattggtggaggaaa ctctaaccacctccttt		AATTT TTAAA	2	YAGINUMA et al. (1987)
3	c3	1	TTTAG AAATC	T A	atatc tatag	TTAAA AATTT	TTTAG AAATG	4	HINO et al. (1989)
4	C29	2	CACGC GTGCG	TG AC	taa att	T A	CCCAG GGGTC	2	NAGAYA et al. (1987)
5	p2	0	GCGCG CGCGC	G C	tggcgcgcgcctgta accgcgcgcggacat		GTCCC CAGGG	1	NAGAYA et al. (1987)
6	pY	2	TCCCC AGGGG		aaattcctcttg tttaaggagaac	TAA ATT	TTGAT AACTA	4	NAGAYA et al. (1987)
7	W93HCC	2	ACAAC TGTTG		ctc gag		AAGAA TTCTT	2	HSU et al. (1988)
8	PLCPRF5-26	1	TTTTT AAAAA	AACG TTGC	tgtagccgggcgtccat acatcggcccgcaggta		GATAC CTATG	3	BERGER and SHAUL (1987)
9	HuH2-2	3	CTGAG GACTC	GGCCTTTGTC CCGGAAACAG			CCAAT GGTTA	2	YAGINUMA et al. (1985)
10	PLCPRF5-14	1	TCACA AGTGT		acagg..... tgtcc......				
					.....gggga .....cccct	AGG TCC	CCAGC GGTCG	1	BERGER and SHAUL (1987)

11	H2-7	3	TCTTC TG AGAAG AC	agctt . . . . . tcgaa.		
				. . . . . .tgta CAACT ATTTT . . . . . .acat GTTGA TAAAA	2	TAKADA et al. (1990)
12	HCY-23	3	GCACT CTAGA CGTGA GATCT	tttaa . . . . . aaatt.		
				. . . . .accac AG GCCAT . . . . .tggtg TC CGGTA	1	TAKADA et al. (1990)
13	HT14NCP	2	CTAGA CA GATCT GT	gaagc. . . . . cttcg		SHIH et al. (1987)
14	W74HCC	2	GACTG AA CTGAC TT	ctgca. . . . . gacgt.		HSU et al. (1988)

Analysis for	Total no. analyzed	No. of positive	Percentage positive
Junctions positive for Topo I site(s)	26	25	96
Junctions positive for multiple Topo I sites	26	19	73

Deletions of large segments of cellular DNA have also been reported in association with HBV integration. One interesting case was that in which HBV integration mediated a deletion of more than 13 kb of cellular DNA at chromosome locus 11p13-14 (ROGLER et al. 1985). Since the Wilms' tumor suppressor gene is located at chromosome locus 11p13 (FRANCKE and RICCARDI 1979), this raised the question as to whether the HBV integration caused deletion of one copy of the Wilms' tumor suppressor gene. As the Wilms' tumor suppressor gene has not yet been cloned, the actual loss of the gene in the tumor awaits further analysis. The possible role of tumor suppressor gene inactivation in HCC is discussed in Sect. 5.5.

4.5 Hepadnavirus Proteins and Hepatocarcinogenesis

Until recently, no biological function had been described for the protein encoded by the X open reading frame of mammalian hepadnaviruses. Due to its nucleotide composition and position in the genome, relative to the position of oncogenes in retroviruses, investigators have suggested that the X-gene was originally a cellular gene (MILLER and ROBINSON 1986). The X-genes of HBV, WHV, and GSHV have since been shown to encode proteins which transactivate transcription from a wide array of target promoters (TWU and SCHLOEMER 1987; SPANDAU and LEE 1988; ZAHM et al. 1988; TWU and ROBINSON 1989; COLGROVE et al. 1989). Transactivation of both hepadnavirus and heterologous promotors has been observed when the X protein was expressed in its native state during productive infection (COLGROVE et al. 1989). Northern blot analysis of reporter gene RNA levels and transcription run on analyses indicate that the *trans*-activation occurs at the level of primary transcription.

Computer analysis of X protein sequences predict an X polypeptide which is a small, soluble, intracellular protein without a signal sequence or other obvious identifying sequence motifs (COLGROVE et al. 1989). Its transactivation activity varies to a great degree in different cell lines and is usually the greatest in liver-derived cell lines (TWU and ROBINSON 1989). However, large differences in trans-activation occur among different hepatoma cell lines using the same target promoter (TWU and ROBINSON 1989). These results suggest that X protein trans-activation requires the participation of other cellular factors, and a recent report has shown that cellular transcription factors Ap1, Ap2, and Sp1 can mediate some X protein transactivation effects (SETO et al. 1990). Indeed, X protein has recently been shown to be a novel protein kinase (T.T. WU and W.S. ROBINSON 1990).

Since the 3′ end of the X-gene lies within the cohesive overlap, integration of hepadnavirus DNA frequently occurs within the X ORF and results in truncation of the X protein coding sequence. In several of these cases, the X proteins expressed are fusion proteins with cellular domains. These X-cellular fusion proteins, which are expressed in HCC, have been shown to maintain their trans-activation potential (WOLLERSHEIM et al. 1988).

Human T-cell leukemia virus (HTLV-1) and bovine leukosis virus (BLV) also contain transactivating factors and appear to participate in oncogenesis in a manner similar to hepadnaviruses. Neither HTLV-1 nor BLV encode their own oncogenes or integrate their proviral DNA in a site-specific manner in tumors (WONG-STAAL and GALLO 1985). However, both encode proteins responsible for transcriptional activation of viral gene expression (MARUYAMA et al. 1987). In addition, the HTLV-1 transactivator gene activates transcription of the interleukin-2 receptor gene (GREENE et al. 1986), suggesting that it plays a role in stimulating lymphoid cell proliferation. In adenovirus, the transcriptional transactivator protein coded by the E1a region has been shown to potentiate the transforming effects of viral and cellular oncogenes (RULEY 1983).

The first evidence directly linking X-gene expression with pathologic changes in the liver has recently been obtained through the use of HBV-X-gene transgenic mice. Two lines of these mice were developed by Dr. GILBERT JAY by injecting mouse oocytes with a HBV DNA fragment comprising the HBV enhancer, X promoter, X ORF, and HBV polyadenylation signal. X-gene expression in all three lines caused a lesion which began as small focal areas of hepatocytes described as clear cells, due to the presence of clear vacuoles which contained large amounts of glycogen. These lesions eventually progressed to involve the whole liver by the age of 1 year, and eventually clear cell adenomas developed (Dr. GILBERT JAY, personal communication). These findings which directly link the HBV-X-gene with pathological changes which occur in the early stages of hepatocarcinogenesis are in contrast to data from other laboratories in which HBV- X-gene transgenic mice did not develop a pathological reaction (LEE et al. 1990). They also demonstrate that the X protein is not a complete oncogene and that additional genetic change is required for continued progression to HCC.

A second transcriptional transactivator has most recently been identified in the preS2/S region of an integrated HBV DNA which was cloned from a hepatocellular tumor (KEHULE et al. 1990). The transactivation activity was produced by a 3′-truncated preS2/S region fused to cellular DNA which encoded a 102 aa viral-cellular fusion protein. A frame shift mutation within the preS2 ORF of the fusion protein completely eliminated the transactivation activity, indicating that the protein itself was responsible for the effects on transcription. An artificially 3′-truncated preS2/S gene from a wild-type HBV genome also exhibited transactivation activity while the intact HBV preS2/S gene was not active as a transactivator. Thus, the cellular sequences were not necessary for the transactivation effect. Studies of the steady state level of reporter gene mRNA showed that the transactivation effect occurred at the transcriptional level.

Since the HuH-4 cell line, from which the original pS2/S transactivator was cloned, strongly expressed the c-*myc* gene, an experiment was conducted to determine whether the pS2/S transactivator could transactivate c-*myc* promoters linked to a reporter gene. The experiments demonstrated that it was capable of transactivating the P2 but not the P_1 promoter of the c-*myc* gene. Interestingly, truncated preS2/S genes are present in many of the integrations which have been cloned from HCC (SHAUL et al. 1984; OGSTON et al. 1982), and a second integration containing a similar 3′-truncated preS2/S sequence also exhibited transactivation activity. These findings support the idea that accidental 3′-truncations of integrated HBV preS2/S genes may frequently contribute to hepatocellular transformation, possibly through the activation of cellular protooncogenes.

5 Studies on the Activities of Protooncogenes, Growth Factors, Tumor Suppressor Genes, Immune Surveillance, Cell–Cell Interactions, Cellular Differentiation, etc. During Carcinogenesis

5.1 General Considerations Regarding the Complexity of Experimental Carcinogenesis

The title of this section was purposely composed to emphasize the complexity of the subject of hepatocarcinogenesis. The biological effects of specific protooncogenes, growth factors, and tumor suppressor genes are undoubtedly affected by factors which vary with the cell type and phase of the cell cycle. Although the predominant cell type in the liver is the hepatocyte, others such as endothelial cells, Kupffer's cells, and inflammatory cells (especially in the case of viral infection) comprise a significant portion (~10%–20%) of the liver mass. In various precancerous states, significant variations in cell populations occur. For example, oval cell proliferation precedes and leads to the formation of precancerous nodules, and endothelial cell proliferation can lead to fibrosis and cirrhosis. Fibroblasts produce a family of growth factors (fibroblast growth factors, FGFs) which induce angiogenesis, and these factors are most certainly necessary for the vascular development of tumors (BURGESS and MACIAG 1989). Their effects on hepatocytes and tumor growth are unknown. Furthermore, inflammatory cells produce a series of interleukins, and platelets produce hepatocyte growth factor (HGF), which is a potent stimulator of hepatocyte growth (NAKAMURA et al. 1989).

Our understanding of the specific cellular and genetic controls which are altered in initiated hepatocytes in precancerous nodules or in transformed hepatocytes is very incomplete. In fact, there is currently no consensus as to the specific protooncogene activities associated with any stage of hepatocarcinogenesis in vivo. Intense study has shown that protooncogene expression can be controlled by gene amplification, gene rearrangement, point mutations, altered transcription (up or down and alternate splicing), and translational and posttranslational mechanisms (FRIEND et al. 1989). At the present time, no single protooncogene or growth factor has been studied thoroughly enough to rule out or establish its universal participation in hepatic oncogenesis.

Tumor cells typically carry multiple genetic changes that act synergistically to produce the full range of neoplastic traits (FRIEND et al. 1989). The requirement of multiple oncogenes is quite consistent with multistep carcinogenesis in the liver. Each step through which a normal cell proceeds on the route to malignancy may be demarcated by the acquisition of one mutant gene or another. Mating experiments with transgenic mice containing liver-targeted oncogenes have clearly shown that oncogenes function synergistically together to induce HCC (SANDGREN et al. 1989). The problem remains which and how many oncogenes are activated or suppressed during naturally occurring HCC. Also, do the activities of oncogenes which participate in the early stages of oncogenesis continue to be required after the cells have progressed further toward malignancy? In some cases, specific oncogenes involved in the premalignant phases of carcinogenesis have been proposed, as is the case for the *ras* oncogenes in premalignant papillomas on mouse skin (BALMAN and PRAGNELL 1983) and the evolution of benign colon polyps into frank carcinomas in the colon (FORRESTER et al. 1987). In the absence of a consensus as to oncogenes and growth factors involved at different stages in hepatic oncogenesis, it is only possible at present to cite examples in which specific oncogenes and growth factors have been implicated in the oncogenic process in liver. A complete understanding of the complex interactions involved in hepatocellular transformation awaits further investigations.

5.2 The *myc* Family of Protooncogenes

One study of c-*myc* expression during liver regeneration, after partial hepatectomy, has shown that c-*myc* mRNA accumulates in a transient manner within the 1st h after surgery followed by a second and also transient activation 8 h later (THOMPSON et al. 1986). Another study showed that c-*myc* mRNA remained elevated during the entire regeneration response (SOBCZAK et al. 1989). Transcription of c-*myc* was studied in a set of related liver epithelial lines (hepatocyte-like oval cells) which exhibited untransformed, partially transformed, or fully transformed phenotypes (BRAUN et al. 1989). The c-*myc* transcription was cell cycle regulated in the untransformed and partially transformed cell lines, similar to its regulation during liver regeneration. However, after continuous passage in vitro the cells acquired the capacity to form tumors, and at this point, expression of c-*myc* was found to be constitutive (BRAUN et al. 1989). Constitutive expression of c-*myc* also occurs in the HepG2 hepatoma cell line, and the accumulation of c-*myc* mRNA is not a consequence of the increased stability of the message (HUBER and THORGEIRSON 1987). Thus, deregulation of c-*myc* transcription may be a general characteristic of hepatocellular tumor cells. As stated earlier, WHV integration also appears to play a direct role in *myc* gene activation in virus carriers (Sect. 4.3).

Studies which have utilized diet as the carcinogenic regimen in rats have supported a role for the c-*myc* gene in precancerous as well as carcinogenic changes in the liver (CHANDAR et al. 1989). When rats were fed a choline-devoid (CD) diet for 12 months followed by a choline-supplemented diet, 73% of the animals developed moderately well-differentiated HCC of trabecular adenomatous or mixed phenotypes. Amplification of the c-*myc* gene was found in 13 of the 13 tumors. The amplification ranged from 2- to 70-fold and was accompanied by increases in c-*myc* gene expression. In addition, gene amplification was also apparent in nontumorous regions of tumor-bearing livers. The authors proposed that gene amplification occurs at many loci (endogenous retroviral loci were also rearranged) during the CD diet and that cells containing c-*myc* amplifications have a selective advantage. A two- to five fold increase in the c-*myc* steady state mRNA level occurred in both livers and tumors of tissues with c-*myc* amplifications. This modest increase appears to provide enough additional c-*myc* expression to deregulate cell proliferation.

The involvement of c-*myc* in hepatocarcinogenesis is also supported by studies with transgenic mice in which a c-*myc* gene was placed under the control of an albumin promotor (SANDGREN 1989). These animals develop mildly hyperplastic lesions in the liver which progress to form benign adenomas. Thus, c-*myc* initiates limited cell regeneration in the context of differentiated hepatocytes in an intact liver. When c-*myc* transgenic mice were mated with transgenic mice containing an activated c-H-*ras* gene, the transgenic littermates which contained both the activated c-*myc* and c-H-*ras* genes developed HCC at a very rapid rate compared with littermates containing either oncogene alone (SANDGREN 1989). However, the HCC appeared as clonal growths protruding from the nontumorous, hyperplastic liver. Therefore, even in these mice additional genetic events were clearly required for hepatocyte transformation, and identification of these additional genetic changes is one of the greatest challenges in tumor biology today.

5.3 *ras*, *raf*, and Other Protooncogenes Activated During Hepatocarcinogenesis

The utilization of standard transfection assays to search for activated cellular protooncogenes in HCC have yielded mostly negative results. This is especially true for studies which have utilized fibroblast cell lines to assay for protooncogenes in the *ras* family. Some reports have, however, detected oncogenes in HBV-associated HCC (OCHIYDA et al. 1986; TAKADA and KOIKE 1989). A problem with this approach has been that the assays are limited to the detection of dominantly transforming genes which can function in both fibroblasts in vitro and hepatocytes in vivo. Recently, a liver epithelial cell line, with oval cell markers, was transformed by transfection with the EJ *ras* oncogene, and the transformed cells gave rise to HCC when injected into nude mice (BRAUN et al. 1987). The nude mouse tumor exhibited a wide spectrum of phenotypes from anaplastic, poorly differentiated HCC to moderately well differentiated HCC.

Such differences would be expected to occur from transfection of a pluripotent oval cell line, and further work in this area may be of value.

The complexity of this issue of hepatocyte transformation is perhaps most easily seen in studies involving chemical carcinogenesis, which have clearly shown that tumor formation varies with and is dependent on the species and strain of rodent employed. Thus, host variation may significantly affect the progress to HCC in hepadnavirus infections. B6C3F1 mice, for instance, have a high spontaneous rate of HCC which is thought to be due to the loss of tumor suppressor genes (as yet unidentified) in these strains. Consistent with this, investigators have been able to detect oncogenes in HCC DNA of B6C3F1 mice. For example, some reports have detected oncogenes in up to 80% of spontaneously occurring HCC (REYNOLDS et al. 1986, 1987). The majority of the oncogenes activated were c-H-*ras* genes with a lesion in codon 61. Activation of the H-*ras* gene was dominant compared with the N-*ras* (0/13) and K-*ras* (2/13) genes. A broad spectrum of activating point mutations in the H-*ras* gene was observed. Interestingly, foci induced by mutations in H-*ras* codons 13 and 117 were slower to appear and less morphologically transformed than foci induced by codon 12 or codon 61 H-*ras* mutants. This could in part account for why these mutations were not previously detected in HCC. Novel, weakly activating point mutations in *raf* genes and four non-*ras* transforming genes were also observed in the experiments (REYNOLDS et al. 1987). Activated oncogenes were also detected in 30% of hepatocellular adenomas in B6C3F1 mice, suggesting that oncogene activation may confer a selective advantage on premalignant hepatocytes (REYNOLDS et al. 1986). Forty percent of the mouse hepatocellular adenomas contained some morphologic features of malignancy, indicating that they are precursors of HCC. Thus, mutations in the *ras* family of protooncogenes (preferably H-*ras*) are capable of expressing themselves in high frequency in the B6C3F1 genetic background and may have a role in precancerous or cancerous changes in some but not all hepadnavirus patients.

5.4 Possible Roles of Paracrine and Autocrine Growth Factors such as TGF-α, TGF-β, HGF, and IGF-II in Hepatocarcinogenesis

A complex array of growth factors and their receptors have been identified which affect the growth and differentiation of numerous cell types. In most cases, these growth factors act through autocrine or paracrine mechanisms to regulate the growth of target cells. Since growth factor- and/or growth factor receptor-induced polyclonal proliferation of cancer precursor cells are important aspects of the etiology of several types of tumors, it is very likely that endogenous liver growth factors also play significant roles in the etiology of HCC.

Current knowledge of the activity of endogenous liver growth factors during liver development or progression to HCC is very fragmentary. The understanding which has been gained has come primarily through studies of growth factor expression during liver regeneration and through analysis of growth factor expression in precancerous liver and HCC. Four endogenous liver growth factors which vary significantly in their activity under different experimental regimens are tumor growth factor alpha (TGF-α), tumor growth factor beta (TGF-B), hepatocyte growth factor (HGF), and insulin-like growth factor II (IGF-II). A brief discussion of some of the evidence linking them to liver cell and tumor growth are presented.

5.4.1 Tumor Growth Factor Alpha

TGF-α is a 50-aa peptide closely related to epidermal factor (EGF). It binds to the same receptor as EGF on hepatocyte plasma membranes and in general has similar biological effects. Increases in TGF-α mRNA in hepatocytes coincide with the peak of hepatocyte DNA synthesis between 24 and 72 h after partial hepatectomy (MEAD and FAUSTO 1989). Addition of TGF-α 20 ng/ml to hepatocyte cultures stimulates approximately a 13-fold increase in DNA synthesis, and this increase is inhibited when TGF-β is added at 100 pg/ml. TGF-α protein is doubled in regenerating liver 24 h after partial hepatectomy, and cultured hepatocytes can secrete TGF-α.

EGF receptors are present on hepatocyte membranes after partial hepatectomy, and some reports have detected modest reductions in receptor mRNAs and proteins. It is difficult to implicate EGF as the ligand responsible for down-regulation of its own receptor during liver regeneration since it is not made in the liver, and its serum concentration does not change after partial hepatectomy (RALL et al. 1985). In contrast, the above results implicate TGF-α as the physiological ligand that participates in an autocrine loop with EGF receptors. TGF-α does not appear to initiate the regenerative response but comes into play later, probably at a time when hepatocytes have already made the transition from G_0 to G_1 and have entered the cell cycle (MEAD and FAUSTO 1989). In spite of its clear role in liver regeneration, no studies have yet shown its expression to be altered (up or down) in either precancerous livers or HCC.

The mature form of TGF-α is cleaved from an integral membrane precursor. The absence of an appropriate protease or protease cleavage site would ensure that the growth factor remains anchored to the plasma membrane. A recent study demonstrated that TGF-α precursor molecules, anchored in the cell surface, can bind to the EGF receptor on adjacent cells, leading to signal transduction (WONG et al. 1989). Such molecules continued to stimulate activity of the EGF receptor kinase on adjacent cells (WONG et al. 1989). It was proposed that membrane-anchored growth factors may lead to prolonged biological responses relative to those of secreted growth factors. Whether such activities are involved in liver cell proliferation and hepatocellular carcinoma is unknown; however, these results point out the limitations of conclusions based solely on mRNA steady state accumulation levels.

5.4.2 Tumor Growth Factor Beta

Partial hepatectomy and the consequent rise and fall of liver regeneration also provide an opportunity to study factors that terminate hepatocyte replication. TGF-β is a potent inhibitor of DNA synthesis in primary monolayer cultures of adult rat hepatocytes (BRAUN et al. 1986). TGF-β $5 \times 10^{-12} M$ also induced a 50% inhibition of EGF-mediated DNA synthesis in primary hepatocytes (BRAUN et al. 1986). Specific saturable receptors for TGF-β were found on normal rat hepatocytes, but specific binding could not be detected on hepatocytes isolated from 24-h regenerating liver (BRAUN et al. 1986). These studies in conjunction with others which show that TGF-β is produced by endothelial cells after partial hepatectomy (THOMPSON et al. 1986) suggest that TGF-β inhibits hepatocyte growth by a paracrine mechanism in the late stages of the partial hepatectomy response.

TGF-β also exerts vastly different effects on the proliferation of normal rat liver epithelial cells versus transformed liver cells. TGF-β inhibited colony formation (83%) and DNA synthesis of immortalized rat liver epithelial cells and primary hepatocytes at levels as low as 20 pg/ml (McMAHON et al. 1986). Transformation of the immortalized liver epithelial cells by treatment with aflatoxin in vitro was associated with a loss of sensitivity of the cells to inhibition by TGF-β. TGF-β inhibition of normal liver epithelial cells caused morphological changes characterized by cytoplasmic hypertrophy and formation of abundant lysosomal derivatives, resembling lipofuscin. One possibility in this context is that TGF-β promotes differentiated functions in these cells.

In HCC of WHV-carrier woodchucks, TGF-β mRNA accumulation occurs only in isolated megakaryocytes (YANG and ROGLER, unpublished data). In SV40-induced HCC in transgenic mice, fibroblast proliferation surrounds and fills up regions of necrosis. TGB-β mRNA accumulation occurs in the regions of fibroblast growth and not in the HCC of these animals (P. SCHIRMACHER and C.E. ROGLER, unpublished data). In these cases, hepatocellular tumor cells may have lost their sensitivity to TGF-β, as occurred in the in vitro transformed liver epithelial cells described above.

5.4.3 Hepatocyte Growth Factor

It has long been known that a serum growth factor is present in rats after partial hepatectomy and other types of chemically induced liver damage. Recently, HGF was purified to homogeneity, and the purified protein was used to clone a full length cDNA (NAKAMURA et al. 1989). HGF is produced in and released from platelets which accumulate in the liver as a result of various types of liver injury. Purified preparations of HGF are potent stimulators of DNA synthesis in primary hepatocytes (NAKAMURA et al. 1986). Studies of the mechanism of this paracrine induction of hepatocyte growth are in their infancy. The initial report of the cloning of HGF cDNA stated but did not present proof that HGF is also produced by hepatocytes. In any event, platelets accumulate in the liver of hepadnavirus cell cycle regulation (DECAPRIO et al. 1988). Thus, a major function of SV40 T Ag is likely to involve alteration of the activity of tumor suppressor genes.

When SV40 T Ag expression is combined with the overexpression of a cellular protooncogene, such as the H-*ras* gene in transgenic mice, HCC develops much more rapidly in the mice with both oncogenes than in those with each oncogene (SANDGREN et al. 1989). Similar to the c-*myc* plus H-*ras* mice, the tumors arising in the SV40 T Ag plus H-*ras* mice do not involve the whole liver and, therefore, arise through the selection of cells which have undergone additional genetic or epigenetic alterations. In addition to the presence of two activated oncogenes, almost all the HCC appearing in SV40 T Ag transgenic mouse liver expresses high levels of IGF-II (P. SCHIRMACHER and C.E. ROGLER, unpublished data), raising the possibility that derepression of IGF-II expression is one of the additional events required for progression to HCC. Interestingly, in humans, the IGF-II gene is located at chromosome 11p15 and the Wilms' tumor suppressor gene is located at chromosome 11p13. This suggests that loss of a suppressor gene at 11p13 could lead to derepression of a growth factor gene, such as IGF-II, at a distant site on the same chromosome. These studies are just beginning to uncover the complex interactions which mediate the activities of different classes of oncogenes.

References

Balman A, Pragnell IB (1983) Mouse skin carcinomas induced *in vivo* by chemical carcinogens have a transforming Harvey-ras oncogene. Nature 303: 72–74

Berger I, Shaul Y (1987) Integration of hepatitis B virus: Analysis of unoccupied sites. J Virol 61: 1180–1186

Beasley RP, Lin CC, Hwang LY, Chien CS (1981) Hepatocellular carcinoma and hepatitis B virus. A perspective study of 22,707 men in Taiwan. Lancet 2: 1129–1133

Braun A, Carr BI, Hayashi I, Branum EL, Moses HL (1986) Inhibition of DNA synthesis in rat hepatocytes by platelet-derived type B transforming growth factor type B. Proc Natl Acad Sci USA 85: 1539–1543

Braun L, Goyette M, Yaswen P, Thompson NL, Fausto N (1987) Growth in culture and turmorigenicity after transfection with the ras oncogene of liver epithelial cells from carcinogen treated rats. Cancer Res 47: 4116–4124

Braun L, Mikumo R, Fausto N (1989) Production of hepatocellular carcinoma by oval cells: cell cycle expression of c-myc and p53 at different stages of oval cell transformation. Cancer Res 49: 1554–1561

Brechot CM, Pourcel C, Louise A, Rain B, Tiollais P (1980) Presence of integrated hepatitis B virus DNA sequences in cellular DNA in human hepatocellular carcinoma. Nature 286: 533–535

Brechot CM, Hadchouel M, scotto J, Fonck M, Potet F, Vyas GN, Tiollais P (1981) State of hepatitis B virus DNA in hepatocytes of patients with hepatitis B surface antigen-positive and- negative liver disease. Proc Natl Acad Sci USA 78: 3906–3910

Brill SJ, DiNardo S, Voelkel-Meiman K, Sternglantz R (1987) Need for DNA topoisomerase I activity as a swivel for DNA replication and for transcription of ribosomal RNA. Nature 326: 414–416

Burgess WH, Maciag T (1989) The heparin-binding (fibroblast) growth factor family of proteins. Annu Rev Biochem 58: 575–606

Cariani E, Lasserre C, Seurin D, Hamelin B, Kemeny F, Franco D, Czech MP, Ullrich A, Brechot C (1988) Differential expression of insulin-like growth factor II mRNA in human primary liver cancers, benign liver tumors and liver tumors and liver cirrhosis. Cancer Res 48: 6844–6849

Chakraborty PR, Ruiz–Opazo N, Shouval D, Shafritz DA (1980) Identification of integrated hepatitis B virus DNA and expression of viral RNA in and HBsAg-producing human hepatocellular carcinoma cell line. Nature 286: 531–533

Champoux JJ (1977) strand breakage by the DNA untwisting enzyme results in the covalent attachment of the enzyme to DNA. Proc Natl Acad Sci USA 74: 3800–3804

Champoux JJ (1978) Mechanism of the reaction catalyzed by the DNA untwisting enzyme: attachment of the enzyme to the 3′-terminus of the nicked DNA. J Mol Biol 118: 441–446

Champoux JJ (1988) Topoisomerase I is preferentially associated with isolated replicating simian virus 40 molecules after treatment of infected cells with camptothecin. J Virol 62: 3675–3683

Champoux JJ, Bullock PA (1988) A possible role for the eucaryotic type I topoisomerase intellegitimate recombination. In: Kucherlapati R, Smith G (eds) Genetic recombination, American Society for Microbiology, Washington, pp 655–656

Chandar N, Lombardi B, Locker J (1989) C-myc gene amplification during hepatocarcinogenesis by a choline-devoid diet. Proc Natl Acad Sci USA 86: 2703–2707

Chen DS, Hoyer BH, Nelson J, Pourcel RH, Gerin JL (1982) Detection and properties of hepatitis B viral DNA in liver tissues from patients with hepatocellular carcinoma. Hepatology 2: 42S–46S

Chisari FV, Ferrari C, Mondelli MU (1989a) Hepatitis B virus structure and biology-mini review. Microbiol Pathog 6: 311–325

Chisari FV, Filippi P, Buras J, McLachlan A, Popper H, Pinkert CA, Palmiter RD, and Brinster RL (1987) Structural and pathological effects of synthesis of hepatitis B virus large envelope polypeptide in transgenic mice. Proc Natl Acad Sci USA 84: 6909–6913

Chisari FV Klopchin K, Moriyama T, Pasquinelli C, Dunsford HA, Sell S, Pinkert CA, Brinster RL, Palmiter RD (1989b) Molecular pathogenesis of hepatocellular carcinoma in hepatitis B virus transgenic mice. Cell 59: 1145–1156

Colgrove R, Simon G, Ganem D (1989) Transcriptional activation of homologous and heterologous genes by the hepatitis B virus gene product in cells permissive for viral replication. J Virol 63: 4019–4026

DeCaprio J, Ludlow JW, Figge J, Shew JY, Huang CM, Lee WH, Marsilio E, Paucha E, Livingston DM (1988) SV40 large tumor antigen forms a specific complex with the product of the retinoblastoma susceptibility gene. Cell 54: 275–283

Dejean A, Sonigo P, Wain–Hobson S, Tiollais P (1984) Specific hepatitis B virus integration in hepatocelular carcinoma DNA through a viral 11-base-pair direct repeat. Proc Natl Acad Sci USA 81: 5350–5354

Dejean A, Bougueleret L, Grzeschik K-H, Tiollais P (1986) Hepatitis B virus DNA integration in a sequence homologous tov-erb-A and steroid receptor genes in a hepatocellular carcinoma. Nature 322: 70–72

DeThe H, Marchio A, Tiollais P (1987) A novel steroid thyroid hormone receptor-related gene in appropriately expressed in human hepatocellular carcinoma. Nature 330: 667–670

Draetta G, Beach D (1988) Activation of cdc2 protein kinase during mitosis in human cells: Cell cycle-dependent phosphorylation and subunit rearrangement. Cell 54: 17–26

Draetta G, Luca F, Westendorf J, Brizuela L, Ruderman J, Beach D (1989) cdc2 protein kinase is complexed with both cyclin a and b: evidence for proteolytic inactivation of mpf. Cell 56: 829–838

Dulak NC, Temin HM (1973) Multiplication-stimulating activity of chicken embryo fibroblasts from rat liver cell conditioned medium: a family of small polypeptides. J Cell Physiol 82: 161–170

Dunsford HA, Sell S (1989) Production of monoclonal antibodies to preneoplastic liver cell populations induced by chemical carcinogens in rats and to transplantable Morris hepatoma. Cancer Res 49: 4984–4900

Dunsford HA, Karnasuta C, Hunt JM, Sell S (1989) Different lineages of chemically induced hepatocellular carcinoma in rats defined by monoclonal antibodies. Cancer Res 49: 4894–4900

Edman JC, Gray P, Valenzuela P, Rall LB, Rutter WJ (1980) Integration of hepatitis B virus sequences and their expression in a human hepatoma cell. Nature 286: 535–538

Evarts RP, Nagy P, Marsden E, Thorgeirson S (1987) A precursor-product relationship exists between oval cells and hepatocytes in rat liver. Carcinogenesis 8: 1737–1740

Farber E (1973) Carcinogenesis-cellular evaluation as a unifying thread: presidential address. Cancer Res 33: 2537–2550

Farber E (1980) Sequential analysis of liver cancer induction. Biochim Biophys Acta 605: 149–166

Farber E (1984) Pre-carcerous steps in carcinogenesis: their physiological adaptive nature. Biochim Biophys Acta 738: 171–180

Forrester K, Almoguera C, Han K, Grizzle WE, Perucho M (1987) Detection of high incidence of K-ras oncogenes during human colon tumorigenesis. Nature 327: 298–303

Wu J, Zhou Z-Y, Judd A et al. (1990) The hepatitis B virus-encoded transcriptional activator Hbx appears to be a novel protein serine/threonine kinase. Cell 63: 687–695

Yaginuma K, Kobayashi M, Yoshida E, Koike K (1985) Hepatitis B virus integration in hepatocellular carcinoma DNA: Duplication of cellular flanking sequences at the integration site. Proc Natl Acad Sci USA 82: 4458–4462

Yaginuma K, Kobayashi H, Kobayashi M, Morishima T, Matsuyama K, Koike K (1987) Multiple integration site of hepatitis B virus DNA in hepatocellular carcinoma and chronic active hepatitis tissues from children. J Virol 61: 1808–1813

Yang DY and Rogler CE (1991) Insulin-like growth factor (IGF-II) and mannose-6-phosphate/IGF-II receptor expression in neoplastic nodules and hepatocellular carcinomas: A potential autocrine role in malignant progression in the liver. Cancer Research (in press)

Yokosuka O, Omata M, Zhou YZ, Imazeki F, Okuda T (1985) Duck hepatitis B virus DNA in liver and serum of chinese ducks: integration of viral DNA in a hepatocellular carcinoma. Proc Natl Acad Sci USA 82: 5180–5184

Zahm P, Hofschneider PH, Koshy R (1988) The HBV X-ORF encodes a transactivator: a potential factor in viral hepatocarcinogenesis. Oncogene 3: 169–177

Human Hepatitis Delta Virus

J. M. TAYLOR

Institute for Cancer Research, Fox Chase Cancer Center, 7701 Burholme Avenue, Philadelphia, PA 19111, USA

Current Topics in Microbiology and Immunology, Vol. 168

of what others have demonstrated for the genomes of certain of the subviral agents of plants (SÄNGER 1987). From both hybridization studies and nucleotide comparisons it would seem that there is no significant sequence relatedness between the hepadnaviruses and HDV (WANG et al. 1987). Also, there is no proven relatedness between the HDV and a host genome.

Inside infected cells it is possible to detect not only the HDV genome but also relatively lower amounts of a complementary RNA, which is called the "antigenome." From the nucleotide sequences of both the genome and antigenome it is possible to predict several potential open reading frames for HDV proteins (WANG et al. 1986, 1987; MAKINO et al. 1987; KUO et al. 1988a). However, at this time, the only one that is known to function during infection is that encoding the delta antigen (WEINER et al. 1988; CHANG et al. 1988). This reading frame, located on the antigenomic RNA (see Fig. 1), is expressed via translation of a specific mRNA that is considerably shorter than the antigenome (see Sect. 4.3 and Fig. 2). In early studies of HDV, only genomic RNA was detected within virions, and antigenomic RNA was estimated to be at least 100-fold less abundant (CHEN et al. 1986). However, by using the sensitivity of the polymerase chain reaction, it has become possible recently to detect antigenomic RNA in purified virions (G. LUC and J. TAYLOR, unpublished observations).

3 Genome Replication

At the outset it should be made clear that whereas hepadnaviruses replicate via reverse transcription of an RNA intermediate (SUMMERS and MASON 1982), HDV does not (CHEN et al. 1986). As will be explained, HDV replicates via RNA-directed RNA synthesis. We will begin with a summary of the experimental evidence and then consider specific models that may explain the replication strategy of this HDV genome.

3.1 Experimental Systems

3.1.1 Infected Animals

As mentioned in Sect. 1.3, HDV infection occurs naturally in humans, and experimental transmission has been achieved in chimpanzees and woodchucks. In all such studies, HDV has been found in association with a hepadnavirus, be it HBV for primates or WHV for woodchucks. One might thus expect that HDV would also replicate in ducklings infected with duck hepatitis B virus (DHBV). However, it is puzzling that contrary to such an expectation and in spite of preliminary results (PONZETTO et al. 1987), several laboratories have been unsuccessful in achieving this goal.

With the three available host systems (humans, chimpanzees, and woodchucks), it has been possible to study not only the replication of HDV but also the interactions between HDV and the helper hepadnavirus, as discussed in Sect. 3.5. In the experimental superinfections of chimpanzees and woodchucks it has been clearly shown that the peak of HDV viremia occurs about 5–7 weeks postinoculation with HDV (PONZETTO et al. 1984). This peak arises and promptly disappears, all within several days. This can be a major experimental problem unless one is able to take multiple serum and even biopsy samples, in order to catch the infection at its peak.

3.1.2 Infected Hepatocytes

TAYLOR et al. (1987b) and later CHOI et al. (1988) have demonstrated that primary cultures of woodchuck hepatocytes are susceptible to infection by HDV that had been grown in a woodchuck that was chronically infected with WHV, and SUREAU et al. (1991) have shown that primary cultures of chimpanzee hepatocytes can also support HDV replication.

Experiments with woodchuck hepatocytes reveal that the cultured cells do not need a prior infection with WHV in order to synthesize HDV RNA replicative intermediates, suggesting that WHV does not contribute to this process (TAYLOR et al. 1987b). This result supports the hypothesis that the only role of a hepadnavirus in HDV replication is to provide the envelope of the mature virion. Curiously, in vivo, coinfection by HDV causes significant suppression of hepadnavirus replication (see Sect. 7).

The results of in vitro infection of primary hepatocyte cultures have so far confirmed the host specificity reported for infected animals. They have also been valuable for in situ hybridization to localize sites of HDV RNA accumulation and for studying other aspects of the time course of HDV infections (TAYLOR et al. 1987b). Using the same system PETCU et al. (1988) and later CHOI et al. (1989) have shown that suramin acts on both HDV and HBV and blocks virus entry. CHOI et al. (1989) have also studied the effects of α-amanatin, acyclovir, and ribavarin. However, neither animal nor tissue culture studies have so far clarified one of the major questions concerning HDV replication, namely, the nature of the polymerase involved in HDV RNA synthesis (see Sects. 3.1.6, 3.8, and 8). There is the possibility that some of the effects of acyclovir and ribavarin as reported by CHOI et al. may have been indirect, acting on the survival of the hepatocyte cultures rather than on HDV replication.

3.1.3 Cells Transfected with cDNA

It has been possible to assemble a full-length cDNA clone of the HDV genome and to insert multimers of this clone into a eukaryotic expression vector. KUO et al. (1989) inserted a trimer of HDV into pSVL so that genomic HDV RNA is expressed via DNA-dependent transcription directed by the late promotor of SV40 to yield a trimer of genomic HDV RNA. Transfection of cell lines with this

3.3.2 Introns

It has been pointed out previously by DINTER-GOTTLIEB (1986) and HADIDI (1986) that there may be some evolutionary relationship between certain of the plant agents and group I introns. More recently DIENER (1989) has even speculated the reverse, that the introns may have evolved from agents like viroids. Do these analogies have any relevance in our attempts to understand HDV? At this point the answer is not clear, but it is interesting, as explained in Sect. 3.6, that the self-cleavage of antigenomic HDV RNA shows some similarities to the removal of introns from mRNA precursors.

3.3.3 *7S* L RNA

Recently there has been a burst of interest as to the possible relationship between the 7S RNA component of the signal recognition particle and certain of the viroids and even HDV. HAAS et al. (1989) first pointed out a patch of sequence homology between certain viroids and 7*S* L RNA. They and later SYMONS (1989a) noted that the homology was in the region of the viroid sequence that is considered to be a determinant of pathogenicity. NEGRO et al. (1989) and YOUNG and HICKE (1990) have claimed a homology between an important region on the HDV genome and the 7*S* L RNA. The region of homology partially overlaps the site on the HDV genome at which in vitro self-cleavage has been shown to occur (as discussed in Sect. 3.6). This has led to speculations regarding the mechanism of pathogenicity of HDV, for example, that HDV might be able to reduce the amount of functional signal recognition particles via cleavage of 7*S* L RNA. At present this proposal is highly speculative. Moreover, doubts can be raised about the significance of the original sequence comparisons (TAYLOR et al. 1990b). The answers will have to come from direct experimental evaluation.

3.4 Rolling-Circle Model

The concept of a rolling-circle model was introduced to explain the replication of viruses that were thought to have a circular intermediate in the synthesis of progeny nucleic acid (BROWN and MARTIN 1965). BRANCH and ROBERTSON (1984) applied such models to explain the replication of certain of the pathogenic plant RNAs; they even invoked a single- and double-circle model. SÄNGER (1987) has reviewed this and three related models for viroid replication. We have invoked another adaptation to explain the replication of HDV (TAYLOR et al. 1989). Recently (TAYLOR 1990b), a more cautious approach to this model has been advocated, largely for the reason that the model contains many highly speculative areas.

3.5 Role of the Hepadnavirus

From the cDNA transfection studies of KUO et al. (1989) and the liposome studies of GLENN et al. (1990), as discussed in Sect. 3.1.3, it is clear that no HBV function or sequence is needed to obtain HDV genome reproduction. However, it is necessary to keep clear the distinction between the full cycle of HDV replication and the more limited concept of genome replication. We are tempted to conclude that the functions of the hepadnavirus are not needed for genome replication, and that the hepadnavirus is only needed for the assembly of new virions and later for reinfection into new host cells. This conclusion may need an additional qualification: It remains possible that in the liver of an infected animal even HDV RNA synthesis may depend upon one or more functions provided a hepadnavirus. A precedent for this distinction between tissue culture and animal studies is seen, for example, in the case of the thymidine kinase gene of herpes simpex virus, which is apparently not needed for virus replication in cultured cells but which in vivo seems to function as a facilitator of virus replication (FIELD and WILDY 1978).

3.6 Self-Cleavage

one of the major predictions of the analogy between HDV and the plant agents was that HDV RNA might demonstrate an in vitro ability to self-cleave. This was shown by SHARMEEN et al. (1988) and subsequently confirmed by WU et al. (1989) and by BRANCH et al. (1991). There is one site of self-cleavage on the HDV genomic RNA and one on the antigenomic RNA (see Fig. 1). These two sites are very similar in sequence, consistent with the extensive intramolecular pairing of the HDV genome. The cleavage reaction is dependent upon a divalent metal ion, such as magnesium. The reaction is also dependent upon temperature. The reaction is a transesterification, producing a 5′-OH and a cyclic 2′,3-monophosphate. KUO et al. (1988b) have elucidated the minimum amounts of contiguous sequence essential for self-cleavage; only about 100 bases are needed, with the cleavage site located near to the 5′-terminus.

For HDV, the self-cleavage of the antigenomic strand is particularly intriguing. The reason is because of the close proximity of several important *cis*-acting sequences, as indicated on Fig. 1. Consider the following: 56 bases 3′ of the termination codon for the delta antigen (see Sect. 4.3) is a polyadenylation signal, AAUAAA. This site is essential for the in vivo expression of the mRNA for the delta antigen (HSIEH et al. 1990). Some 20 bases further 3′ is the site at which poly(A) is added to make this mRNA. Then, 34 bases further 3′ is the self-cleavage site, which KUO et al. (1988b) have shown depends for its in vitro activity on 3′ and 5′ flanking sequences. Not only is there this close-packing, but the 16 bases 5′ of the self-cleavage site are all pyrimidines, analogous to the intron-exon border of mRNA introns. And, for the total of 34 bases between the adenylation site and the self-cleavage site there is only one adenosine, making the adenylation site somewhat like the branch point of a mRNA intron (SMITH et al. 1989).

With the self-cleaving RNAs of plants it has been possible not only to define the domain necessary for self-cleavage (BUZAYUN et al. 1986; FORSTER and SYMONS 1987; PRODY et al. 1986) but also to go further and separate out a catalytic domain from a substrate domain (UHLENBECK 1987). Such studies of HDV RNA are being carried out by several laboratories. To date, the RNA structure that induces self-cleavage of HDV has not been defined (LAI et al. 1991; BELINSKY and DINTER-GOTTLIEB 1991). As reviewed by SYMONS (1989b), amongst those RNAs of plant agents that do undergo self-cleavage, two types of structures have been defined: a so-called "hammer-head," which is found in many of these agents, and a structure so far unique to one strand of the satellite of tobacco ringspot virus (and a closely related virus). The HDV genome and antigenome cleavage structures do not seem to belong to either of these classes.

One puzzling aspect of the in vitro studies of KUO et al. (1988b) was that while a minimum sequence capable of undergoing self-cleavage could be defined, a slightly longer sequence was not functional. One such extra sequence was that which allowed the cleavage region to fold, intramolecularly, into part of the predicted unbranched rod structure. This inhibition could also be achieved intermolecularly, by the presence, in addition to a short molecule that was competent for cleavage, of a second RNA that allowed the formation by intermolecular base pairing, with this short RNA of a structure analogous to the unbranched rod structure. The data that the rod structure was inhibitory to self-cleavage suggested the corollary, that the rod structure might be an inducer of self-ligation, as discussed in Sect. 3.7.

3.7 Self-Ligation

SHARMEEN et al. (1989) sought to detect self-ligation on HDV RNA for three reasons. (1) There was already a precedent with the plant agents; (2) the model, introduced in Sect. 3.6, of the rod structure as an inducer of self-ligation seemed plausible; and (3) USHER and MCHALE (1976) had already described what proved to be another example of self-ligation. Their work had been concerned with obtaining protein-free ligation of hexamers of adenylic acid. These hexamers, like the cleavage products of HDV RNA, contained a 5′-hydroxyl and a 2′,3′-cyclic monophosphate. The hexamers could be induced to undergo efficient self-ligation by the presence of a backbone molecule, polyuridylic acid, that served to align the hexamers head-to-tail. Therefore, SHARMEEN et al. took two small fragments derived by self-cleavage of antigenomic HDV RNA, aligned these on an RNA from the other side of the rod structure, and using the conditions defined by USHER and MCHALE, obtained an efficient self-ligation reaction. One puzzling aspects of this work was that at most half of the self-ligated molecules could undergo a subsequent cleavage. It is possible that the other half had ligated not by a 3′–5′ linkage but by a nonbiological 2′–5′ linkage.

WU and LAI (1989) have also observed a self-ligation. They found that as long as the two self-cleavage products were allowed to remain associated after cleavage then ligation could be achieved simply by the removal of the magnesium ions.

3.8 RNA-Directed RNA Synthesis

As discussed above, the available data suggest that HDV, just like the plant agents, is replicated by RNA-directed RNA synthesis. Two important questions follow: where does this synthesis take place, and how is it achieved? As implied in Sect. 3.2, there is the suggestion that the location is the nucleolus, consistent with the findings for one of the plant viroids (HARDERS et al. 1989). The polymerase(s) involved has yet to be identified. It probably is not virus-coded, considering that the delta antigen is, so far, the only known viral protein and, at 22 kDa, is probably too small to act alone as an RNA polymerase. Thus, viral RNA synthesis probably requires a host polymerase. As known host polymerases normally use DNA rather than RNA templates, there will have to be a redirection. As reviewed by SYMONS et al. (1985), there are many claims for RNA polymerases which can be redirected in vitro to copy viroid RNAs. However, as SYMONS et al. point out, the transcription must first be understood in a more biological setting. There is only one report of experiments on viroid transcription in isolated nuclei; SPIESMACHER et al. (1985) claimed that DNA-dependent RNA polymerases I and II were used for the synthesis of genomic and antigenomic RNAs, respectively. Currently, such nuclear run-on assays are being applied to HDV in an attempt to determine which polymerase is actually redirected.

An important related question is the location on the HDV genome and antigenome of the sites of initiation of RNA-directed RNA synthesis. One preliminary piece of information is that the antigenomic RNA that acts as mRNA for the delta antigen (Sect. 4.3) appears to start 5 bases away from the end of the intramolecular rod structure (HSIEH et al. 1990). We do not know whether initiation of the opposite strand, the genomic RNA, occurs at one or multiple sites.

Yet another problem is to determine the *cis*-acting sequence features that are used to specify the initiation of genomic and antigenomic RNA transcription. Maybe the most direct way to solve this will be from a study of the sequences of HDV minivariants (Sect. 6) that are replication competent. It is also possible that a simple examination of the sequences conserved on such minivariants might help us to deduce the sequences essential for initiation of RNA synthesis.

contained RNA sequence variants. This is, of course, true for most other viruses, including the plant agents (KEESE et al. 1988). With the sequencing of other HDV isolates (MAKINO et al. 1987), it became clear that there were not only base substitutions but also insertions and deletions relative to the HDVs sequenced by WANG et al. (1986). As mentioned earlier (Sect. 1), there are HDV outbreaks that produce unusually severe liver disease, and from the sequence of the HDV RNA obtained from cases, it may be possible to recognize genome feature(s) essential for an increase in pathogenicity. Such studies have been attempted with the plant viroids, as reviewed by KEESE et al. (1988), and there are precedents for as little as a single base change being a determinant of pathogenicity. For both the plant agents and for HDV it is quite possible that different, more or less pathogenic strains, arise in several ways.

Another kind of HDV variant has recently been observed in the work of LUO et al. (1990) and SUREAU et al. (1989). This is a variant that arises both with high efficiency and specificity during HDV genome replication within a single cell, in the absence of further cell-to-cell spread. It was observed that when cultured cells were transfected with a cDNA clone of HDV, there was not only the initiation of HDV genome replication but also the accumulation of genomes with a specific sequence alteration. This alteration also occurred in a chimpanzee transfected by HDV cDNA. The alteration was in the amber termination codon for the 195-aa delta antigen, thereby directing the coding of the amino acid tryptophan and creating a delta antigen that was 19 aa longer, for a total of 214 aa, i.e., the large delta antigen. This modification only occurred in transfection experiments when there was RNA-dependent RNA synthesis, and in fact, the extent of modification depended on the extent of genome synthesis. LUO et al. have speculated that this change might involve RNA unwindase activity. This is a deaminase activity which is present in most animal cells (BASS and WEINTRAUB 1988; WAGNER and NISHIKURA 1988), and has recently been implicated in the modification of the genomes of negative-strand viruses (BASS et al. 1989) and in specific developmental control of mRNA species (KIMELMAN and KIRSCHNER 1989).

LUO et al. also showed by transfection experiments that the large form of the delta antigen produced as a result of this sequence alteration, unlike the small form, was unable to support HDV genome replication. They thus speculated that the larger form might in some way reduce the extent of HDV replication, leading in the short term to a milder acute disease or in the long term to an increased chance of progressing to chronicity.

Yet another class of HDV variants are those with genomes that are significantly smaller than the normal size of about 1700 b. These are called minivariants (TAYLOR et al. 1990a). Such species, with lengths down to about 400 b, have been detected in the livers of infected chimpanzees and especially infected woodchucks. Apparently the molecules lack the *cis*-acting sequences needed for packaging, because they are not found in virions. However, the observation that they are produced in relatively significant amounts implies that they must have the *cis*-acting elements required for replication. Thus, it will be informative to isolate and sequence these species.

7 Interaction with the Hepadnavirus

Of course, the major interaction of HDV with HBV occurs during packaging. Presumably there are *cis*-cting signals both on the HDV RNAs and transacting signals on the HBV proteins and*or on the delta antigen that regulate this interaction. At present, however, nothing is known about the regulating signals.

Earlier (Sect. 3.5), it was shown that HDV RNA synthesis does not depend upon the hepadnavirus. However, as summarized below, in the biological setting of an in vivo infection, the situation is still quite complex:

1. The hepadnavirus does provide the coat, but it is not yet clear whether this coat is exactly the same as present on the hepadnaviruses. Some data of BONINO et al. (1986) argue for a difference in the relative amounts of the various related forms of HBV surface antigen in HDV as compared with HBV.
2. There are data that HDV infection can, at least transiently, have a suppressive effect on hepadnavirus replication (e.g., CHEN et al. 1988 and SUREAU et al. 1989).
3. From what we now know, a significant part of an in vivo HDV infection may be nonproductive, that is, resulting from the spread of HDV particles to liver cells (and perhaps to nonliver cells) that are not already infected by HBV nor likely to be infected. Such a situation might occur, for example, following HDV infection of a surface antigen carrier that is no longer viremic. There are actual immunofluorescence microscopy and in situ hybridization data from GOWANS et al. (1988) to prove such events occur in natural infections. Such nonproductive infection may in theory have significant overall consequences for the host. These consequences may be deleterious or as speculated below may in some situations even be advantageous.
4. A number of investigators, working with HDV-infected patients experiencing progressive liver failure, have as a last resort offered a liver transplantation to their patients. It has been noted that such patients can promptly establish HDV RNA synthesis in the new liver (as distinct from a full replication cycle with packaging of new virions). The spread of delta to the new liver seems in turn to give this organ enhanced resistance to infection by HBV, which in these studies was also being produced prior to the transplantation (GRIDELLI et al. 1991). This could be the explanation for the better prognosis for a liver transplant when the patient has a prior HDV infection, relative to patients infected with HBV alone.

8 Future Directions

Throughout this review there has been an attempt to discuss questions regarding HDV and its replication that are currently accessible to experimental analysis. Prominent examples include the mechanism of RNA-directed RNA synthesis, the development of an in vivo system for genome replication, the regulation of the mRNA for the delta antigen, and the roles of the delta antigen. Another important problem is the mechanism of packaging. We have suggested that there might be *cis*-acting sequences on the HDV genome, but at the same time, it might be possible that other viruses could provide the proteins that would allow for packaging of HDV into infectious particles. It is possible to speculate that other subviral agents analogous to HDV will be found to be associated with human diseases (TAYLOR 1990 b).

A particularly important direction for the future will be to apply the information we have obtained about HDV replication to problems of human disease. For example, we have with HDV the wherewithal to deliver an RNA that can, with minimal help, be amplified 300 000-fold in a single cell. If the HDV genome were modified to contain antisense (WEINTRAUB 1990) or antigene (HAZELOFF and GERLACH 1988) activities, one might achieve the ability to inactivate specific genes efficiently. With an appropriate delivery system, one might be able to ameliorate certain chronic infections, such as HBV, which is normally a major risk factor for the development of hepatocellular carcinoma (BEASLEY 1988).

Acknowledgments. Financial support was from grants CA-06927, RR-05539, and AI-26522 from the National Institutes of Health, grant MV-7M from the American Cancer Society, and an appropriation from the Commonwealth of Pennsylvania. Discussions of unpublished work were with Mei Chao, Sen-Yung Hsieh, Guangxiang Luo, and Wang-Shick Ryu as well as with Jeff Glenn. Valuable comments on the manuscript were given by Bill Mason, Allison Jilbert, and Tom London.

References

Banerjee AK (1987) Transcription and replication of rhabdoviruses. Bacteriol Rev 55: 599–629

Bass BL, Weintraub H (1988) An unwinding activity that covalently modifies its double-stranded RNA substrate. Cell 55: 1089–1098

Bass BL, Weintraub H, Cattaneo R, Billeter MA (1989) Biased hypermutation of viral RNA genomes could be due to unwinding/modification of double-stranded RNA. Cell 56: 331

Beasley RP (1988) Hepatitis B virus: the major etiology of hepatocellular carcinoma. In: Fortner JG, Rhoads JE (eds) Accomplishments in cancer research: 1987 prize year. Lippincott, Philadelphia pp 80–106

Belinsky M, Dinter-Gottlieb G (1991) Characterizing the self-cleavage of a 135 nucleotide ribozyme from genomic hepatitis delta virus. In: Gerin JL, Purcell RH, Rizzeto M (eds) The hepatitis delta virus. Liss, New York, pp. 265–274

Bergmann KF, Gerin JL (1986) Antigens of hepatitis delta virus in the liver and serum of humans and animals. J Infect Dis 154: 702–705

Birnstiel ML, Schaufele FJ (1989) Structure and function of minor snRNPs. In: Birnstie ML (ed) Small ribonucleoprotein particles. Springer, Berlin Heidelberg New York, pp 155–182

Bonino F, Hoyer W, Shih JW-K, Rizzetto M, Purcell RH, Gerin JL (1984) Delta hepatitis agent: structural and antigenic properties of the delta-associated particles. Infect Immun 43: 1000–1005

Bonino F, Heermann K-H, Rizzetto M, Gerlich WH (1986) Hepatitis delta virus: protein composition of delta antigen and its hepatitis B virus-derived envelope. J Virol 58: 945–950

Bonino F, Negro F, Baldi M, Brunetto M, Chiaberge E, Capalbo M, Maran E, Lavarini C, Rocca N, Rocca G (1987) The natural history of chronic delta hepatitis. In: Rizzetto M, Gerin JL, Purcell RH (eds). The hepatitis delta virus and its infection. Liss New York, pp 145–152

Bonino F, Negro F, Brunetto MR, Verme G (1990) Hepatitis delta infection. Prog Liver Dis 9: 485–496

Branch AD, Robertson HD (1984) A replication cycle for viroids and small infectious RNAs. Science 223: 450–455

Branch AD, Levine BJ, Baroudy BM, Buckler-White A, Gerin JL, Robertson HD (1991) The novel tertiary structure in delta RNA may function as a ribozyme control element. In: Gerin JL, Purcell RH, Rizzeto M (eds) The hepatitis delta virus. Liss, New York, pp. 257–264

Brown F, Martin SJ (1965) A new model for virus ribonucleic acid replication. Nature 208: 861–863

Buitrago B, Popper H, Hadler SC, Thung SN, Gerber MA, Purcell RH, Maynard JE (1986) Specific histologic features of Santa Marta hepatitis: a severe form of hepatitis δ-virus infection in northern South America. Hepatology 6: 1285–1291

Buzayun JM, Gerlach WL, Bruening G (1986) Non-enzymatic cleavage and ligation of RNAs complementary to a plant virus satellite RNA. Nature 323: 349–353

Chang M-F, Baker SC, Soe LH, Kamahora T, Keck JG, Makino S, Govindarajan S, Lai MMC (1988) Human hepatitis delta antigen is a nuclear phosphoprotein with RNA binding activity. J Virol 62: 2403–2410

Chao M, Hsieh S-Y, Luo G, Taylor J (1991) The antigen of human hepatitis delta virus: the significance of the two major electrophoretic forms. In: Gerin JL, Purcell RH, Rizzeto M (eds) The hepatitis delta virus. Liss, New York, pp. 257–281

Chen P-J, Kalpana G, Goldberg J, Mason W, Werner B, Gerin J, Taylor J (1986) The structure and replication of the genome of the hepatitis delta virus. Proc Natl Acad Sci USA 83: 8774–8778

Chen P-J, Chen D-S, Chen C-R, Chen Y-Y, Chen H-MH, Lai M-Y, Sung J-L (1988) δ infection in asymtomatic carriers of hepatitis B surface antigen: low prevalence of δ activity and effective suppression of hepatitis B virus replication. Hepatology 8: 1121–1124

Choi S-S, Rasshofer R, Roggendorf M (1988) Propagation of woodchuck hepatitis delta virus in primary woodchuck hepatocytes. Virology 167: 451–457

Choi S-S, Rasshofer R, Roggendorf M (1989) Inhibition of hepatitis delta virus RNA replication in primary woodchuck hepatocytes. Antiviral Res 12: 213–222

Davis GL, Balart LA, Schiff ER, Lindsay K, Bodenheimer HC, Prrillo RP, Carey W, Jacobson IM, Payne J, Dienstag JL, Van Thiel DH, Tamburro C, Lefkowitch J, Albrecht J, Meschievitz C, Ortego TJ, Gibas A, and the Hepatitis International Therapy Group (1989) Treatment of chronic hepatitis C with recombinant interferon alpha: a multicenter randomized, controlled trial. N Eng J Med 321: 1501–1510

Diener TO (1989) Circular RNAS: relics of precellular evolution? Proc Natl Acad Sci USA 86: 9370–9374

Diener TO, Prusiner SB (1985) The recognition of subviral pathogens. In: Marmorosch K, McKelvey JJ Jr (eds) Subviral pathogens of plants and animals: viroids and prions. Academic, Orlando, pp 3–20

Dinter-Gottlieb G (1986) Viroids and virusoids are related to group I introns. Proc Natl Acad Sci USA 83: 6250–6254

Ellens H, Doxey S, Glenn JS, White JM (1989) Delivery of macromolecules into cells expressing a viral membrane fusion protein. Methods Cell Biol 31: 155–176

Field HJ, Wildy P (1978) The pathogenicity of thymidine kinase-deficient mutants of herpes simplex virus in mice. J Hyg (Lond) 81: 267–277

Forster AJ, Symons RH (1987) Self-cleavage of viroid RNA is performed by the proposed 55-nucleotide active site. Cell 50: 9–16

Ganem D, Varmus HE (1987) The molecular biology of hepatitis B viruses. Annu Rev Biochem 56: 651–693

Glenn JS, Taylor JM, White JM (1990) In vitro synthesized hepatitis delta virus RNA initiates genome replication in cultured cells. J Virol 64: 3104–3107

Gowans EJ, Baroudy BM, Negro F, Ponzetto A, Purcell RH, Gerin JL (1988) Evidence for the replication of hepatitis delta virus RNA in nuclei after in vivo infection. Virology 167: 274–278

Gridelli B, Grendele M, Colledan M, Rossi G, Fassati LR, Feria G, Maggi U, Reggiani P, Doglia M, Lunghi G, Galmarini D (1991) Liver transplantation (LTx) and delta virus (DV) infection. In: Gerin JL, Purcell RH, Rizzeto M (eds) The Hepatitis Delta Virus, Liss, New York, p. 471

Haas B, Klanner A, Ramm K, Sänger HL (1989) The 7S RNA from tomato leaf tissue resembles a signal recognition particle RNA and exhibits a remarkable sequence complementarity to viroids. EMBO J 7: 4063–4074

Hadidi A (1986) Relationships of viroids and certain other plant pathogenic nucleic acids to group I and II introns. Plant Mol Biol 7: 129–142 highly phosphorylated C terminus that is essential for replication but not for RNA packaging. J Virol 63: 2995–3000

Sculliard GR, Pollard RB, Smith JL, Sacks SL, Gregory PB, Robinson WS, Merigan TC (1981) Antiviral treatment of chronic hepatitis B virus infection. I. Changes in viral markers with interferon combined with adenine arabinoside. J Infect Dis 143: 772–783

Seeger C, Ganem D, Varmus HE (1986) Biochemical and genetic evidence for the hepatitis B virus replication strategy. Science 232: 477–484

Sells MA, Chen ML, Acs G (1987) Production of hepatitis B virus particles in HepG2 cells transfected with cloned hepatitis B virus DNA. Proc Natl Acad Sci USA 84: 1005–1009

Sherker AH, Hirota K, Omata M, Okuda K (1986) Foscarnet decreases serum and liver duck hepatitis B virus DNA in chronically infected ducks. Gastroenterology 91: 818–824

Smee DF, Knight SS, Duke AE, Robinson WS, Matthews TR, Marion PL (1985) Activities of arabinosyladenine monophosphate and 9-(1,3-dihydroxy-2-propoxymethyl) guanine against ground squirrel hepatitis virus in vivo as determined by reduction in serum virion-associated DNA polymerase. Antimicrob Agents Chemother 27: 277–279

Smith CC, Aurelian L, Reddy MP, Miller PS, Ts'o POP (1986) Antiviral effect of an oligo (nucleoside methylphosphonate) complementary to the splice junction of herpes simplex virus type 1 immediate early pre-mRNAs 4 and 5. Proc Natl Acad Sci USA 83: 2787–2791

Stein CA, Cohen JS (1988) Oligodeoxynucleotides as inhibitors of gene expression: a review. Cancer Res 48: 2659–2668

Su TS, Lai CJ, Huang JL, Lin LH, Yauk YK, Chang CM, Lo SJ, Han SH (1989) Hepatitis B virus transcript produced by RNA splicing. J Virol 63: 4011–4018

Sureau C, Romet-Lemonne J-L, Mullins JI, Essex M (1986) Production of hepatitis B virus by a differentiated human hepatoma cell line after transfection with cloned circular HBV DNA. Cell 47: 37–47

Tsiquaye KN, Collins P, Zuckerman AJ (1986) Antiviral activity of the polybasic anion, suramin and acyclovir in hepadna virus infection. J Antimicrob Chemother 18: 223–228

Tuttleman JS, Pourcel C, Summers J (1986) Formation of the pool of covalently closed circular viral DNA in hepadnavirus-infected cells. Cell 47: 451–460

Tuttleman JS, Pugh JC, Summers JW (1986) In vitro experimental infection of primary duck hepatocyte cultures with duck hepatitis B virus. J Virol 58: 17–25

Twu JS, Robinson WS (1989) Hepatitis B virus X gene can transactivate heterologous viral sequences. Proc Natl Acad Sci USA 86: 2046–2050

Ueda K, Tsurimoto T, Nagahata T, Chisaka O, Matsubara K (1989) An in vitro system for screening anti-hepatitis B virus drugs. Virology 169: 213–216

Venkateswaran PS, Millman I, Blumberg BS (1987) Effects of an extract from Phyllanthus niruri on hepatitis B and woodchuck hepatitis viruses: in vitro and in vivo studies. Proc Natl Acad Sci USA 84: 274–278

Viola LA, Barrison IG, Coleman JC, Paradinas FJ, Murray-Lyon IM (1981) The HBe antigen-antibody system and its relationship to clinical and laboratory findings in 100 chronic HBsAg carriers in Great Britain. J Med Virol 8: 169–175

Will H, Reiser W, Weimer T, Pfaff E, Buscher M, Sprengel R, Cattaneo R, Schaller H (1987) Replication strategy of human hepatitis B virus. J Virol 61: 904–911

Wu T-T, Coates L, Aldrich CE, Summers J, Mason WS (1990) In hepatocytes infected with duck hepatitis B virus, the template for viral RNA synthesis is amplified by an intracellular pathway. Virology 175 (in press)

Yaginuma K, Shirakata Y, Kobayashi M, Koike K (1987) Hepatitis B virus (HBV) particles are produced in a cell culture system by transient expression of transfected HBV DNA. Proc Natl Acad Sci USA 84: 2678–2682

Zerial A, Thuong NT, Hélène C (1987) Selective inhibition of the cytopathic effect of type A influenza viruses by oligodeoxynucleotides covalently linked to an intercalating agent. Nucleic Acids Res 15: 9909–9919

Zuckerman AJ (1987) Screening of antiviral drugs for hepadna virus infection in Pekin ducks: a review. J Virol Methods 17: 119–126

Inherited Hepatic Enzyme Defects as Candidates for Liver-Directed Gene Therapy

A. L. HORWICH

There are a small number of inherited hepatic defects that are so devastating that virtually 100% of the affected infants either die or suffer severe developmental disability. For these defects, serious consideration must be given to any new treatment that offers the possibility of altering the extremely poor prognosis. In the desparate clinical setting of these single gene defects, two forms of therapy have recently been considered—hepatic transplantation and gene therapy. Here, I discuss two severe single gene disorders that might be candidates for such novel therapies, outline their natural histories, elaborate on the diagnostic tools available, and describe the result of hepatic transplantation in one case. I conclude that there remains a place for the development of cell or gene therapy for these severe disorders, speculate on potential strategies, and suggest that, if a safe and effective strategy could be developed, other less severe conditions of the liver could ultimately be considered for such therapy.

Howard Hughes Medical Institute and Department of Human Genetics, Yale University School of Medicine, New Haven, CT 06510, USA

Current Topics in Microbiology and Immunology, Vol. 168

newborns for accumulation of phenylalanine have enabled rapid diagnosis of the condition in the newborn period and effective treatment by restriction of dietary phenylalanine. This has permitted normal development, but the diet is fairly restrictive and in many social settings compliance is poor. This most immediately results in loss of IQ points of the affected child. Thus a definitive therapy would be desirable.

Alpha-1-Antitrypsin Deficiency. In patients with deficiency of this enzyme, inadequate antiprotease activity results in lung disease produced by unchecked activity of elastase and other proteases. Typically such disease arises in middle adult life, particularly in individuals who smoke. Because this enzyme is normally produced by the liver and released into the circulation, it seems that it could just as easily be produced by cell or gene transfer into a nonhepatic site.

Carbohydrate Disorders. Inherited conditions, including several of the glycogen storage diseases, galactosemia, and hereditary fructose intolerance, involve hepatic enzymes and could be corrected by provision of the normal gene to hepatocytes. In the first condition, considering the most common defect, glucose-6-phosphatase deficiency (von Gierke's disease), this would lead to normal glycogen metabolism and would both remove the present need for frequent glucose or corn starch feedings and remove risks of catastrophic hypoglycemia. In the second disorder, it could remove the toxic effects of galactose metabolites that can damage the lens, brain, and ovary. In the third case, it would remove the risk of hepatic or systemic injury from toxic metabolites that accumulate following ingestion of fructose.

Disorders of Amino Acid Metabolism. In addition to the defects described earlier, OTC deficiency, CPS deficiency, and FAH deficiency, other disorders might also be corrected. Propionic acidemia and methylmalonic acidemia are disorders, often severe, that produce acidemia due to accumulation of unusual organic acids behind blocks in the pathway of three carbon metabolism. Similarly, the defect of branch chain amino acid degradation, maple syrup urine disease, might similarly be corrected.

Lipid Metabolism. Familial hypercholesterolemia, due to LDL receptor deficiency, might be a prime candidate as one of the most common dominant mutations of humans—1:500 individuals is a heterozygote. Ability to produce additional normal receptor at the hepatocyte surface would most likely lead to a lowering of plasma LDL in these individuals (Yokode et al. 1990). Other types of defect of lipoprotein metabolism might also be corrected at the liver.

Porphyrin Metabolism. A major class of porphyrias involves the liver. For example, acute intermittent porphyria is the most common such defect. Ability to supply additional porphobilinogen deaminase activity might prevent the abdominal attacks and neurologic disturbances that characterize this disorder.

In summary, there is great therapeutic potential for a strategy that is genuinely safe and efficient. We have entered an era where animals models for many candidate disorders are being produced. These systems can be used to test and further develop the creative approaches that may emerge for this class of disorders.

Acknowledgement. I thank Wayne Fenton and Saul Brusilow for helpful discussions and encouragement during preparation of this manuscript.

References

Arne PH, Houser ER, Thomas GH, Herman G, Hess D, Brusilow S (1990) Episodes of hyperammonemic coma, often postpartum, occurring in women who have a mutation at the OTC locus: New Engl J Med 322: 1652–1655

Batshaw ML, Brusilow S, Waber L, Blom W, Brubakk AM, Burton BK, Can HM, Kerr D, Mamunes P, Matalon R, Myerberg D, Schafer IA (1982) Treatment of inborn errors of urea synthesis: activation of alternative pathways of waste nitrogen synthesis and excretion. N Engl J Med 306: 1387–1392

Brusilow SW, Horwich AL (1989) Urea cycle enzymes. In Scriver CL, Beaudet AL, Sly WS, Valle D (eds) The metabolic basis of inherited disease, 6th edn. McGraw-Hill, New York, pp 629–663

Brusilow SW, Tinker J, Batshaw ML (1980) Amino acid acylation a mechanism of nitrogen excretion in inborn errors of urea synthesis. Science 207: 659–661

Brusilow SW, Danney M, Waber LJ, Batshaw M, Burton B, Levitsky L, Roth K, McKeethren C, Ward J (1984) Treatment of episodic hyperammonemia in children with inborn errors of urea synthesis. N Engl J Med 310: 1630–1634

Demetriou AA, Whiting JF, Feldman D, Levenson SM, Chowdhury NR, Moscion AD, Kram M, Chowdhury JR (1986) Replacement of liver function in rats by transplantation of microcarrier-attached hepatocytes. Science 233: 1190–1192

Fox JE, Rosenberg LE (1988) Toward a molecular understanding of ornithine transcarbamylase deficiency. Adv Neurol 48: 71–81

Fox JE, Hack AM, Fenton WA, Golbus MS, Winter S, Kalousek F, Rozen R, Brusilow SW, Rosenberg LE (1986) Prenatal diagnosis of ornithine transcarbamylase deficiency with use of DNA polymorphisms. N Engl J Med 315: 1205–1208

Grompe M, Muzny DM, Caskey CT (1989) Scanning detection of mutations in human ornithine transcarbamylase by chemical mismatch cleavage. Proc Natl Acad Sci USA 86: 5888–5892

Goldsmith LA, LaBerge C (1989) Tyrosinemia and related disorders. In Scriver CL, Beaudet AL, Sly WS, Valle D (eds) The metabolic basis of inherited disease, 6th edn McGraw-Hill, New York, pp 556–562

Haldane JBS (1935) The rate of spontaneous mutation of a human gene. J Genet 31: 317–326

Hata A, Tsuzuki T, Shimada K, Takiguchi M, Mori M, Matsuda I (1988) Structure of the human ornithine transcarbamylase gene. J Biochem (Tokyo) 103: 302–308

Hata A, Setoyama C, Shimada K, Takeda E, Koruda Y, Akaboshi I, Matsuda I (1989) Ornithine transcarbamylase deficiency resulting from a C to T substitution in exon 5 of the ornithine transcarbamylase gene. Am J Hum Genet 45: 123–127

Hirsch RC, Lavine JE, Chang LJ, Varmus HE, Ganem (1990) Polymerase gene products of hepatitis B viruses are required for genomic RNA packaging as well as for reverse transcription. Nature 344: 552–555

Hodges PE, Rosenberg LE (1989) The spf-ash mouse a missense mutation in the ornithine transcarbamylase gene also causes aberrant mRNA splicing. Proc Natl Acad Sci USA 86: 4142–4146

Holzgreve W, Golbus MS (1984) Prenatal diagnosis of ornithine transcarbamylase deficiency utilizing fetal liver biopsy. Am J Hum Genet 36: 320–328

Horwich AL, Fenton WA, Williams KR, Kalousek F, Kraus JP, Doolittle RF, Konigsberg W, Rosenberg

Current Topics in Microbiology and Immunology

Volumes published since 1985 (and still available)

Vol. 116: **Willis, Dawn B. (Ed.):** Iridoviridae. 1985. 65 figs. X, 173 pp. ISBN 3-540-15172-9

Vol. 122: **Potter, Michael (Ed.):** The BALB/c Mouse. Genetics and Immunology. 1985. 85 figs. XVI, 254 pp. ISBN 3-540-15834-0

Vol. 124: **Briles, David E. (Ed.):** Genetic Control of the Susceptibility to Bacterial Infection. 1986. 19 figs. XII, 175 pp. ISBN 3-540-16238-0

Vol. 125: **Wu, Henry C.; Tai, Phang C. (Ed.):** Protein Secretion and Export in Bacteria. 1986. 34 figs. X, 211 pp. ISBN 3-540-16593-2

Vol. 126: **Fleischer, Bernhard; Reimann, Jörg; Wagner, Hermann (Ed.):** Specificity and Function of Clonally Developing T-Cells. 1986. 60 figs. XV, 316 pp. ISBN 3-540-16501-0

Vol. 127: **Potter, Michael; Nadeau, Joseph H.; Cancro, Michael P. (Ed.):** The Wild Mouse in Immunology. 1986. 119 figs. XVI, 395 pp. ISBN 3-540-16657-2

Vol. 128: 1986. 12 figs. VII, 122 pp. ISBN 3-540-16621-1

Vol. 129: 1986. 43 figs., VII, 215 pp. ISBN 3-540-16834-6

Vol. 130: **Koprowski, Hilary; Melchers, Fritz (Ed.):** Peptides as Immunogens. 1986. 21 figs. X, 86 pp. ISBN 3-540-16892-3

Vol. 131: **Doerfler, Walter; Böhm, Petra (Ed.):** The Molecular Biology of Baculoviruses. 1986. 44 figs. VIII, 169 pp. ISBN 3-540-17073-1

Vol. 132: **Melchers, Fritz; Potter, Michael (Ed.):** Mechanisms in B-Cell Neoplasia. Workshop at the National Cancer, Institute, National Institutes of Health, Bethesda, MD, USA, March 24–26, 1986. 1986. 156 figs. XII, 374 pp. ISBN 3-540-17048-0

Vol. 133: **Oldstone, Michael B. (Ed.):** Arenaviruses. Genes, Proteins, and Expression. 1987. 39 figs. VII, 116 pp. ISBN 3-540-17246-7

Vol. 134: **Oldstone, Michael B. (Ed.):** Arenaviruses. Biology and Immunotherapy. 1987. 33 figs. VII, 242 pp. ISBN 3-540-17322-6

Vol. 135: **Paige, Christopher J.; Gisler, Roland H. (Ed.):** Differentiation of B Lymphocytes. 1987. 25 figs. IX, 150 pp. ISBN 3-540-17470-2

Vol. 136: **Hobom, Gerd; Rott, Rudolf (Ed.):** The Molecular Biology of Bacterial Virus Systems. 1988. 20 figs. VII, 90 pp. ISBN 3-540-18513-5

Vol. 137: **Mock, Beverly; Potter, Michael (Ed.):** Genetics of Immunological Diseases. 1988. 88 figs. XI, 335 pp. ISBN 3-540-19253-0

Vol. 138: **Goebel, Werner (Ed.):** Intracellular Bacteria. 1988. 18 figs. IX, 179 pp. ISBN 3-540-50001-4

Vol. 139: **Clarke, Adrienne E.; Wilson, Ian A. (Ed.):** Carbohydrate-Protein Interaction. 1988. 35 figs. IX, 152 pp. ISBN 3-540-19378-2

Vol. 140: **Podack, Eckhard R. (Ed.):** Cytotoxic Effector Mechanisms. 1989. 24 figs. VIII, 126 pp. ISBN 3-540-50057-X

Vol. 141: **Potter, Michael; Melchers, Fritz (Ed.):** Mechanisms in B-Cell Neoplasia 1988. Workshop at the National Cancer Institute, National Institutes of Health, Bethesda, MD, USA, March 23–25, 1988. 1988. 122 figs. XIV, 340 pp. ISBN 3-540-50212-2

Vol. 142: **Schüpach, Jörg:** Human Retrovirology. Facts and Concepts. 1989. 24 figs. 115 pp. ISBN 3-540-50455-9

Vol. 143: **Haase, Ashley T.; Oldstone Michael B. A. (Ed.):** In Situ Hybridization 1989. 22 figs. XII, 90 pp. ISBN 3-540-50761-2

Vol. 144: **Knippers, Rolf; Levine, A. J. (Ed.):** Transforming Proteins of DNA Tumor Viruses. 1989. 85 figs. XIV, 300 pp. ISBN 3-540-50909-7

Vol. 145: **Oldstone, Michael B. A. (Ed.):** Molecular Mimicry. Cross-Reactivity between Microbes and Host Proteins as a Cause of Autoimmunity. 1989. 28 figs. VII, 141 pp. ISBN 3-540-50929-1

Vol. 146: **Mestecky, Jiri; McGhee, Jerry (Ed.):** New Strategies for Oral Immunization. International Symposium at the University of Alabama at Birmingham and Molecular Engineering Associates, Inc. Birmingham, AL, USA, March 21–22, 1988. 1989. 22 figs. IX, 237 pp. ISBN 3-540-50841-4

Vol. 147: **Vogt, Peter K. (Ed.):** Oncogenes. Selected Reviews. 1989. 8 figs. VII, 172 pp. ISBN 3-540-51050-8

Vol. 148: **Vogt, Peter K. (Ed.):** Oncogenes and Retroviruses. Selected Reviews. 1989. XII, 134 pp. ISBN 3-540-51051-6

Vol. 149: **Shen-Ong, Grace L. C.; Potter, Michael; Copeland, Neal G. (Ed.):** Mechanisms in Myeloid Tumorigenesis. Workshop at the National Cancer Institute, National Institutes of Health, Bethesda, MD, USA, March 22, 1988. 1989. 42 figs. X, 172 pp. ISBN 3-540-50968-2

Vol. 150: **Jann, Klaus; Jann, Barbara (Ed.):** Bacterial Capsules. 1989. 33 figs. XII, 176 pp. ISBN 3-540-51049-4

Vol. 151: **Jann, Klaus; Jann, Barbara (Ed.):** Bacterial Adhesins. 1990. 23 figs. XII, 192 pp. ISBN 3-540-51052-4

Vol. 152: **Bosma, Melvin J.; Phillips, Robert A.; Schuler, Walter (Ed.):** The Scid Mouse. Characterization and Potential Uses. EMBO Workshop held at the Basel Institute for Immunology, Basel, Switzerland, February 20–22, 1989. 1989. 72 figs. XII, 263 pp. ISBN 3-540-51512-7

Vol. 153: **Lambris, John D. (Ed.):** The Third Component of Complement. Chemistry and Biology. 1989. 38 figs. X, 251 pp. ISBN 3-540-51513-5

Vol. 154: **McDougall, James K. (Ed.):** Cytomegaloviruses. 1990. 58 figs. IX, 286 pp. ISBN 3-540-51514-3

Vol. 155: **Kaufmann, Stefan H. E. (Ed.):** T-Cell Paradigms in Parasitic and Bacterial Infections. 1990. 24 figs. IX, 162 pp. ISBN 3-540-51515-1

Vol. 156: **Dyrberg, Thomas (Ed.):** The Role of Viruses and the Immune System in Diabetes Mellitus. 1990. 15 figs. XI, 142 pp. ISBN 3-540-51918-1

Vol. 157: **Swanstrom, Ronald; Vogt, Peter K. (Ed.):** Retroviruses. Strategies of Replication. 1990. 40 figs. XII, 260 pp. ISBN 3-540-51895-9

Vol. 158: **Muzyczka, Nicholas (Ed.):** Viral Expression Vectors. 1991. Approx. 20 figs. Approx. XII, 190 pp. ISBN 3-540-52431-2

Vol. 159: **Gray, David; Sprent, Jonathan (Ed.):** Immunological Memory. 1990. 38 figs. XII, 156 pp. ISBN 3-540-51921-1

Vol. 160: **Oldstone, Michael B. A.; Koprowski, Hilary (Ed.):** Retrovirus Infections of the Nervous System. 1990. 16 figs. XII, 176 pp. ISBN 3-540-51939-4

Vol. 161: **Racaniello, Vincent R. (Ed.):** Picornaviruses. 1990. 12 figs. X, 194 pp. ISBN 3-540-52429-0

Vol. 162: **Roy, Polly; Gorman, Barry M. (Ed.):** Bluetongue Viruses. 1990. 37 figs. X, 200 pp. ISBN 3-540-51922-X

Vol. 163: **Turner, Peter C.; Moyer, Richard W. (Ed.):** Poxviruses. 1990. 23 figs. X, 210 pp. ISBN 3-540-52430-4

Vol. 164: **Bækkeskov, Steinnun; Hansen, Bruno (Ed.):** Human Diabetes. 1990. 9 figs. X, 198 pp. ISBN 3-540-52652-8

Vol. 165: **Bothwell, Mark (Ed.):** Neuronal Growth Factors. 1991. 14 figs. IX, 173 pp. ISBN 3-540-52654-4

Vol. 166: **Potter, Michael; Melchers, Fritz (Ed.):** Mechanisms in B-Cell Neoplasia 1990. 143 figs. XIX, 380 pp. ISBN 3-540-52886-5

Vol. 167: **Kaufmann, Stefan H. E. (Ed.):** Heat Shock Proteins and Immune Response. 1991. 18 figs. IX, 214 pp. ISBN 3-540-52857-1